LENSES ON LEARNING

Supervision: Focusing on Mathematical Thinking

Readings

for May
Reading 10 p. 132 -
(2 questions)

Edited by
Catherine Miles Grant, Barbara Scott Nelson, Amy Shulman Weinberg,
Annette Sassi, Ellen Davidson, Sheila Gay Buzzee Holland

Center for the Development of Teaching
Education Development Center
Newton, Massachusetts

DALE SEYMOUR PUBLICATIONS
Pearson Learning Group

National Science Foundation

This work was supported by the National Science Foundation under Grant No. ESI-9731242 and by The Pew Charitable Trusts. Any opinions, findings, conclusions, or recommendations expressed here are those of the authors and do not necessarily reflect the views of these organizations.

Note: Every effort has been made to locate the copyright owner of material used in this textbook. Omissions brought to our attention will be corrected in subsequent editions.

Art and Design: Jim O'Shea

Editorial: Stephanie P. Cahill

Marketing: Maureen Christensen and Douglas Falk

Production and Manufacturing: Mark Cirillo, Alia Lesser

Publishing Operations: Carolyn Coyle, Richetta Lobban

Copyright © 2006 by the Education Development Center, Inc. Published by Dale Seymour Publications®, an imprint of Pearson Learning Group, a division of Pearson Education, Inc., 299 Jefferson Road, Parsippany, NJ 07054. All rights reserved. No part of this book may be reproduced or transmitted in any form or by any means, electronic or mechanical, including photocopying, recording, or by any information storage and retrieval system, without permission in writing from the publisher. Blackline masters excepted. For information regarding permission(s), write to Rights and Permissions Department.

Dale Seymour Publications® is a registered trademark of Dale Seymour Publications, Inc.

ISBN 0-7652-7029-3

Printed in the United States of America

1 2 3 4 5 6 7 8 9 10   09 08 07 06 05

1-800-321-3106
www.pearsonlearning.com

Lenses on Learning: Supervision: Focusing on Mathematical Thinking

Project Staff, Education Development Center

Catherine Miles Grant

Barbara Scott Nelson

Amy Shulman Weinberg

Annette Sassi

Ellen Davidson

Sheila Gay Buzzee Holland

Glenn Natali

Pilot-Test Facilitators and Sites

Jeffrey Benson
*Principal and Director of Educational Services,
Germaine Lawrence School*
Arlington, MA

Joanne Gurry
*Assistant Superintendent,
Arlington Public Schools*
Arlington, MA

Joseph Petner
Principal, Haggerty School
Cambridge, MA

Debra Shein-Gerson
Elementary Mathematics Curriculum Coordinator
Brookline, MA

These pilot tests took place under the auspices of the Education Collaborative of Greater Boston (EDCO), Lesley College, and the Merrimack Education Center

Field Test Sites

Houston Independent School District
Houston, TX

Clark County School District
Las Vegas, NV

Portland Public Schools
Portland, OR

New York Public Schools Community District 2
New York, NY

The University of Wisconsin at Milwaukee
Milwaukee, WI.

Project Evaluators, Education Development Center

Barbara Miller

Michael Foster

Project Advisors

Diane Briars
Pittsburgh Public Schools

Nancy Dickerson
Boston Public Schools

Judy Mumme
Mathematics Renaissance K–12

Joseph Murphy
The Ohio State University

Mildred Collins Pierce
Harvard Graduate School of Education

Susan Jo Russell
TERC

James Spillane
Northwestern University

Virginia Stimpson
University of Washington

Philip Wagreich and Kathy Kelso
University of Illinois at Chicago

CONTENTS

Introduction 1

Session 1   **Changing Mathematics Classes, Changing Supervision.** 5
     HOMEWORK   Changing Perspectives 6
     READING 1   Changing Perspective in Curriculum and Instruction 7
                         by J. Nolan and P. Francis

Session 2   **Building Intellectual Communities in Mathematics Classrooms** 21
     HOMEWORK   Learning and Teaching Mathematics 22
     READING 2   Learning Mathematics while Teaching 23
                         by S. J. Russell, et al.

Session 3   **Linking Intellectual Community with Mathematical Inquiry** 33
     HOMEWORK   Observing in Classrooms: Building an Intellectual Community 34
                         PRE-OBSERVATION CONFERENCE QUESTIONS 35
                         INTELLECTUAL COMMUNITY OBSERVATION GUIDE 36
     READING 3   What's All This Talk about "Discourse"? 38
                         by D. Ball, S. Friel

Session 4   **Observing for Content: Listening to Children's Ideas About Fractions** 45
     HOMEWORK   Approaches to Solving a Fair-sharing Problem and Magical Hopes 46
     READING 4   Approaches to Solving a Fair-sharing Problem 47
                         by C. Twomey Fosnot and M. Dolk
     READING 5   Fraction Tracks Transcript 53
     READING 6   Magical Hopes: Manipulative and the Reform of Math Education 59
                         by D. Ball

Session 5   **Supporting Generative Learning: How We Talk With Teachers** 69
     HOMEWORK   Rethinking How We Talk with Teachers and Observing for Content... 70
                         PRE-OBSERVATION CONFERENCE QUESTIONS 73
                         MATH CONTENT OBSERVATION GUIDE 74
     READING 7   Capturing Teachers' Generative Change: A Follow-up Study Of Professional Development in Mathematics 77
                         by M. Franke, T. Carpenter, L. Levi, and E, Fenema

Session 6   **Observing How Knowledge is Constructed in Mathematics Classrooms** 111
- HOMEWORK   "Canceling" Zeroes 112
- READING 8   "Canceling" Zeroes Transcript 113

Session 7   **Building Mathematics Understanding: More than the Sharing of Ideas** 123
- HOMEWORK   Supporting Generative Learning 124
  - PRE-OBSERVATION CONFERENCE QUESTIONS 125
  - LEARNING & PEDAGOGY OBSERVATION GUIDE 126
- READING 9   Designing Packages Transcripts 128

Session 8   **Bringing It All Together: Distributing Supervisory Practices and Fostering Generative Growth** 131
- HOMEWORK   Sources of Authority for Supervision 132
- READING 10   Sources of Authority for Supervision 133
  by T. Sergiovanni and R. Starratt

Observation Guides 148

Resource List 157

Introduction to the Course

Standards-based elementary mathematics classrooms, with their emphasis on mathematical thinking and reasoning, pose new challenges for those of you who observe in classrooms with the purpose of either assisting or assessing the teacher. The teaching in these classrooms is based on fundamentally different assumptions about what knowledge is and what teaching and learning entail. According to one perspective, learning is the absorption of knowledge dispensed by the teacher or a text. In such classrooms, observers look for evidence of dispensing and absorption of knowledge in the classroom structures and in teacher and student behaviors (for example, clear explanation of new ideas by the teacher, opportunities for students to apply new ideas or practice new skills, students observably "on task" throughout the lesson). In a standards-based classroom, mathematical thinking itself is the focus of classroom activity. In order to understand what is happening there, observers must look for the central intellectual ideas of the lesson and the classroom structures and practices that provide opportunities for students to develop those mathematical ideas. Observers' purposes in classroom observations are to assess the quality of mathematics teaching and learning and to support the teacher's continued learning. (Of course, at this time of transition many classrooms are mixtures of both modes of instruction and observers will need to be able to identify and understand both.)

This course is designed to help you develop the skills to be effective observers in a standards-based mathematics classroom. Through the course, you will learn to focus on the mathematical essence of a lesson and to engage teachers in productive conversations after observing in their classrooms. You will also reflect on what *you* can gain from observing in standards-based classrooms and how this knowledge can inform your work more broadly.

Thematic Strands

The course is designed around two thematic strands that develop in interaction through the eight sessions. A third important idea, that of distributed supervisory practices, is considered in the final session of the course.

STRAND 1. DEVELOPING AN EYE FOR MATHEMATICS CLASSROOMS

In this strand, you will examine three aspects of classroom observations: the mathematics content of the lesson; the teaching and learning that are taking place; and the nature of the intellectual community in the classroom.

Developing the capacity to discern the features of a classroom that are central for student learning is not an easy task. You will learn to see beyond the physical and behavioral aspects of the classroom in order to grasp the intellectual activity underlying the classroom practices. To support this learning, you will use an observation tool developed especially for this course, called the *Observation Guide*. Through a systematic set of questions, the Observation Guide helps users focus attention on the mathematics, learning

& pedagogy, and intellectual community in a classroom. You will practice using it as you view videotaped clips of mathematics lessons and observe actual mathematics lessons in your own schools or districts.

STRAND 2. RETHINKING ADMINISTRATORS' TALK WITH TEACHERS ABOUT MATHEMATICS, LEARNING, AND TEACHING.

The course is grounded in the belief that, just as students actively construct new knowledge, teachers must be active participants in constructing their own new knowledge about mathematics, learning, and teaching. While this can happen in professional development programs, it can also happen even more powerfully as teachers reflect on events in their own classroom. When teachers reflect on children's mathematical thinking in their own classrooms, they become *generative learners*. That is, their knowledge continues to grow and contributes to an ever-more developed understanding of effective mathematics instruction. In the second strand of the course, you will consider processes that can help support teachers in developing an orientation of curiosity about their students' mathematical thinking.

In the pre- and post-observation conferences that you may have with teachers, you will explore a process we call "collaborative inquiry," in which both the teacher and administrator share their curiosity and questions about student thinking. These questions can prompt informal investigations from which both you and the teacher can construct new knowledge about mathematics, learning, and teaching. In this way, both you and the teacher can benefit from generative learning. Through the various tasks in the course—working with key mathematical ideas in the elementary classroom, seeing how students make sense of these ideas, and learning to use pre- and post-observation conferences for collaborative inquiry—you can begin to see how supervisory practices can support generative learning.

DISTRIBUTED SUPERVISORY PRACTICES

The course focuses largely on the teacher-supervisor dyad, which is likely to be a familiar model for most of you. In the final session, you will step outside of this model to consider a redistribution of the responsibility and authority for classroom observation across a variety of roles in the school. You will explore ways in which classroom observation and teacher supervision can become a context for school-wide professional development for both teachers and administrators.

Mathematics

While this is not a mathematics course for school administrators, mathematics is fundamental to its design. Two central tenets of the course are that mathematics is about a set of *ideas* as well as procedures rather than *solely about procedures* and that observers should focus on these ideas when conducting classroom observations. Thus, you will need to have experience working with essential mathematics concepts from the K–8 curriculum. The mathematics topics we have chosen for this curriculum—patterns and fractions—are key topics in the mathematics curriculum throughout these grades. Pattern work, especially common in the early grades, helps to lay the groundwork for more advanced mathematics such as algebra. Fractions, while an important topic in their own right, are integrally related to work in other areas such decimals, percents, and ratios. We have presented activities for these topics that we hope you will find relevant as well as interesting and challenging.

Viewing Videotaped Classroom Episodes

In almost every session you will be viewing a videotaped clip of children talking about their mathematical thinking or of children and teachers working together in a mathematics classroom. These clips were chosen for the ideas about mathematics, teaching, and learning they make accessible for consideration and the diversity of teachers and students they depict. In watching these videotaped classroom episodes you will learn to focus on the mathematical thinking that is taking place in the classroom and the instruction supporting the students' development of mathematical ideas. Learning to "see" in new ways requires an orientation of curiosity and discernment, not one of evaluation of the quality of the teaching. As you watch the videos you should be asking yourself, "What can be learned bout teaching mathematics from this glimpse of a classroom?"

School-based Observations

Three times in this course you will be asked to do a classroom observation in your own school or district. These observations will include a pre-observation conference at which you will gather information about the class and the lesson you are about to observe. You will use the course's Observation Guide in these observations. Your facilitator will discuss this activity early in the course and suggest that you identify 2 teachers that you can observe regularly. These should be teachers who are interested in changing their instructional practice and who are in an off-year in the evaluation cycle so they can view themselves as partners in your learning.

Reflective Discussion

You and the other administrators in your *Lenses on Learning* class will function as a reflective community, thinking through and talking together about the ideas that emerge from the readings and videotaped clips. Session after session, you will experience for yourself the social and interpersonal characteristics of inquiry- or discourse-based classrooms. You will see all participants working to think through hard and significant issues. Showing respect for other participants' ideas and developing the trust and commitment needed to support each others' thinking are hallmarks of a *Lenses on Learning* course.

Readings

This book of *Readings* contains all of the homework assignments and readings for each of the sessions the course. In addition, a complete set of the Observation Guides can be found at the end of the book as well as a Resource List. Some of the readings, as well as the materials described in the Resource List, are valuable references to share with members of your school community: teachers, parents, and school boards. We also encourage you to make use of the Observation Guides as you work with teachers to make the transition to a standards-based classroom in which both students' mathematical thinking and teachers' generative learning are supported.

SESSION 1

Changing Mathematics Classes, Changing Supervision

As administrators, you may be faced with new challenges when you supervise elementary and middle school teachers who are using a standards-based approach to the teaching of mathematics. Just as teachers have to rethink the purposes and processes of their instructional practices when they adapt a standards-based approach, you may want to rethink some key aspects of your supervisory practice for the standards-based classroom. In this course, we will talk about two ideas that can help you in your supervisory practices:

1) *Reorienting the focus of your observations in the classroom.* We refer to this as "Developing an Eye for Mathematics Classrooms." You will learn to attend to the central ideas in mathematics, teaching, learning, and assessment when you observe teachers and students engaged in the study of mathematics. This means going beyond observing the physical and behavioral elements of the classroom to developing an eye for what is at the heart of the standards-based classroom.

2) *Rethinking your talk with teachers about mathematics, learning, and teaching.* When teachers focus on making sense of their students' mathematical ideas, they set in motion a process, which we refer to as *generative learning*, in which they continually enhance their knowledge about their students' thinking. Such learning can contribute to the ongoing refinement of their teaching practices. In this course you will develop the skills that can support and enhance this kind of learning for teachers. One way to support teachers' generative learning is through "collaborative inquiry." This course focuses on the kind of generative learning that can take place in both teachers and administrators when they transform classroom observation into collaborative inquiry

While teacher supervision is typically seen as a way to help *teachers* learn and to monitor the quality of teaching in the classroom, in this course we propose that it can also be a medium of learning for you as administrators as well.

This first session lays the foundation for the entire course on classroom observation and teacher supervision. We begin with a discussion on changing perspectives on curriculum, instruction, and teacher supervision.

Changing Perspectives

In preparation for the first session of the course Lenses on Learning: Classroom Observation and Teacher Supervision, please do the following:

Read *Changing Perspectives in Curriculum and Instruction,* by Nolan and Francis.

1. Mark those paragraphs or short sections that seem particularly important or thought-provoking, and be prepared to discuss them in class.

2. In addition, choose one of the implications for supervision described on pages 12–18 and quoted below. Drawing on both your own experiences and reflections as a supervisor and/or supervisee, write about why this implication is an important one to consider in the context of reframing the paradigm for teacher supervision.

Implications for supervision articulated by Nolan and Francis:

- Teachers should be viewed as active knowledge constructors of their own knowledge about learning and teaching.

- Supervisors should be viewed as collaborators in creating knowledge about learning and teaching.

- The emphasis on data collection during supervision should change from almost total reliance on paper-and-pencil observation instruments to capture the events of a single period of instruction to the use of a variety of data sources to capture a lesson as it unfolds over several periods of instruction.

- Both general principles and methods of teaching as well as content-specific principles and methods of teaching should be attended to during the supervisory process.

- Supervision should become more group oriented rather than individually oriented.

Implication: _____

READING 1

Changing Perspectives in Curriculum and Instruction* **

James Nolan and Pam Francis

> Come writers and critics, prophesy with your pen
> And keep your eyes wide, the chance won't come again
> But don't speak too soon for the wheel's still in spin
> And there's no tellin' now where she's aimin'
> And the losers now will be later to win
> For the times they are a-changin'
>
> Bob Dylan, "The Times They Are A'Changin," 1964

This chapter considers the potential impact of contemporary theories of learning and teaching on supervisory practice. Although theorists usually conceive of curriculum and instruction as separate entities, we have chosen not to consider the two separately in this chapter. Curriculum and instruction are frequently separated for purposes of discussion and analysis of the educational process; but in the learning-teaching act, decisions about what to teach (i.e., curriculum) and how to teach it (i.e., instruction) must be reconciled and unified. It is in the learning-teaching act that supervision finds its focus and direction.

Educational practices—and indeed all of human behavior—are guided largely by what Sergiovanni (1985) has termed mindscapes. Mindscapes are mental frameworks or paradigms through which we envision reality and our place in reality. They are usually more implicit and unexamined than explicit. As such, mindscapes are taken for granted and provide a set of beliefs or assumptions that exert a tremendous influence on behavior. Sergiovanni states:

> Mindscapes provide us with intellectual and psychological images of the real world and the boundaries and parameters of rationality that help us to make sense of the world. In a very special way, mindscapes are intellectual security

---

*AUTHOR NOTE: We are grateful to Bernard Badiali, J. Robert Coldiron, and Lee Goldsberry for comments on earlier versions of this chapter.

**From *Supervision in Transition: The 1992 ASCD Yearbook*, edited by Carl D. Glickman. "Changing Perspectives in Curriculum and Instruction" by Nolan, James, and Francis, Pam. Alexandria, VA: Association for Supervision and Curriculum Development. Copyright © 1992. ASCD. Reprinted by permission. All rights reserved.

blankets on the one hand, and road maps through an uncertain world on the other (Sergiovanni 1985, p. 5).

The major thesis of this chapter is that the mindscapes that currently drive both supervision theory and practice will undergo significant alteration as a result of important changes in educators' conceptions of learning and teaching that have evolved during the 1980s. We have developed this thesis through a three-part structure: (1) an examination of traditional views of the learning-teaching process, (2) an examination of changing perspectives on the learning-teaching process, and (3) an examination of the implications of these changing perspectives for the practice of supervision.

Traditional Views of Learning and Teaching

The traditional view of the learning-teaching process, which has dominated instruction in most schools, can be captured in five fundamental beliefs about learning. The power of these beliefs rests not in any particular one, but rather in the fact that they constitute a mutually reinforcing system of beliefs. Even though these beliefs are very powerful in driving much of what we currently do in the name of educational practice, for most educators they have remained largely implicit and unexamined. In fact, we derived our descriptions of these beliefs from an analysis of what schools and educators actually do as they attempt to educate learners, rather than from an analysis of what schools and educators espouse. These five fundamental beliefs are:

1. Learning is the process of accumulating bits of information and isolated skills.

2. The teacher's primary responsibility is to transfer his knowledge directly to students.

3. Changing student behavior is the teacher's primary goal.

4. The process of learning and teaching focuses primarily on the interactions between the teacher and individual students.

5. Thinking and learning skills are viewed as transferable across all content areas.

These five beliefs have important implications for teaching. Given these beliefs, the most important teaching tasks are the following:

- Organizing and structuring the learning material in the most appropriate sequence.

- Explaining concepts clearly and unambiguously.

- Using examples and illustrations that can be understood by students.

- Modeling appropriate application of desired skills.

- Structuring and organizing practice sessions with instructional material so that it will be retained more effectively in long-term memory and transferred appropriately to other contexts.

- Assessing student learning by requiring students to reproduce the desired knowledge and skills on paper-and-pencil tests or through other observable means.

These beliefs have resulted in a teacher-centered conception of teaching and supervision in which the teacher's observable behavior during instruction occupies the center stage of the educational drama. The supervisor works one-to-one with each teacher in a two-step process: (1) the supervisor uses paper-and-pencil observation instruments to carefully capture and document the teacher's observable behavior during instruction; and (2) the supervisor and teacher come together in a conference designed primarily to relate the teacher's observable behavior to both individual student behavior and to research findings on generalizable teaching behaviors that seem to be effective in promoting student learning.

Changing Perspectives on Learning and Teaching

During the 1980s, the shape of educational practice slowly began to change, creating a new mindscape about human learning. This new framework has the potential not only to change teaching behavior on a large-scale basis, but also to cause us to fundamentally alter our beliefs about supervision. This new mindscape, or view of learning and teaching, can also be encapsulated in several interrelated beliefs about the nature of learning and teaching. Some of these beliefs are based on theories of learning that are relatively new; others are based on theories of learning that have existed for many years but have exerted little influence on practice.

1. *All learning, except for simple rote memorization, requires the learner to actively construct meaning.* Learners construct meaning by taking new information, relating it to their prior knowledge, and then putting their new understandings to use in reasoning and problem solving. "In this process, each person is continuously checking new information against old rules, revising the old rules when discrepancies appear and reaching new understandings or constructions of reality" (Brooks, 1990, p. 68). For learning to occur, the learner must actively engage in the mental processes necessary to construct the new meanings and understandings. Although learning theorists have held this belief for many years (see Dewey, 1902), only recently have concerted efforts been made to help practitioners put this notion into practice (see Lampert, 1990).

2. *Students' prior understandings of and thoughts about a topic or concept before instruction exert a tremendous influence on what they learn during instruction.* What people learn is never a direct replica of what they have read or been told or even of what they have been drilled on. We know that to understand something is to interpret it and further that an interpretation is based partly on what we've been told or have read but also on what we already know and on general reasoning and logical abilities (Brandt, 1988–89, p. 15).

One of the teacher's most important tasks must be to explore the conceptions that learners bring with them to the classroom and help them achieve a new, more refined understanding of those concepts. When learners' preexisting conceptions are inaccurate, the teacher must provide experiences that assist the learners to recognize the inaccuracies. Otherwise, their misconceptions are not likely to change as a result of instruction. "It is not sufficient to simply present students with the correct facts. One has to change the concepts or schemas that generated the inaccurate beliefs" (Bransford & Vye, 1989, p. 188).

3. *The teacher's primary goal is to generate a change in the learner's cognitive structure or way of viewing and organizing the world.* The most important factor in any learning-teaching situation is not the observable behavior of either the teacher or the learner. The single most important factor in determining how much a student learns during instruction is the learner's cognitive processing of information during instruction (Anderson, 1989). Changes in observable behavior are important because they can be used to infer that the learner's cognitive structure has changed, but changes in behavior are an indicator of learning and a result of learning, not the learning itself.

4. *Because learning is a process of active construction by the learner, the teacher cannot do the work of the learning.* Students must do the work of learning (Schlechty, 1990). The teacher's task is to help learners acquire the skills and dispositions needed to carry out the work of learning. This means: (a) helping learners acquire learning and thinking strategies; (b) helping learners acquire the metacognitive understanding needed to choose the appropriate learning strategy for a given instructional task and to self-monitor the use of the strategy; and (c) motivating learners to engage in appropriate thinking during instruction. The teacher moves from the role of protagonist to that of director or drama coach, and the student becomes the main character in the educational drama.

5. *Learning in cooperation with others is an important source of motivation, support, modeling, and coaching.* In contrast to the traditional view of learning as a solitary process, the new mindscape recognizes the important role that peers can play in the learning process by sharing responsibility for the learning of all group members. Most successful instructional programs designed to teach higher order cognitive skills prescribe the use of cooperative learning groups focused on meaning-construction activities. Such activities provide a type of cognitive apprenticeship in which students have multiple opportunities to observe others do the work that they are expected to do (Resnick and Klopfer, 1989). There is ample evidence that when students are engaged in cooperative learning activities that are structured to include both group interdependence and individual accountability, they learn more (Slavin, 1989-90).

6. *Content-specific learning and thinking strategies play a much more important role in learning than was previously recognized.* Until the past decade, much of the research on learning focused on learning strategies and skills that were general in nature and applied across subject matter. "One of the great luxuries of the old style research on learning was that you could look for principles that had general validity. Now we believe that we must first immerse ourselves in the study of how people learn particular things in particular environments" (Brandt, 1988–89, p. 14). In the past few years, the pendulum has swung from an exclusive emphasis on general thinking and learning skills to an increasing emphasis on content-specific learning and thinking skills. Perkins and Salomon (1989) agree with the contention that content-specific learning skills were neglected by educational researchers for a long period of time and suggest that learning and thinking

skills are most likely a synthesis of general cognitive strategies and context or content-specific techniques.

As was true for research on learning, process-product research on teaching (which provides the basis for much of our current work in supervision and staff development) has focused almost exclusively on teaching techniques that are applicable across grade levels and subject matter. This heavy emphasis on general principles and methods of teaching to the exclusion of content-specific principles and methods has come under fire from a number of educational researchers in recent years.

One content specialist, Henry (1986), argues that the field of instructional supervision and its emphasis on general notions of teaching has violated the field of English education through the institutionalization of behaviorist views of learning and teaching. Henry paints a picture of thousands of English teachers scurrying to write behavioral objectives, create improved feedback and management loops, and use mastery learning strategies. He sees these activities as antithetical to the very nature of English.

> What is neglected or generally omitted is the fundamental probing of instruction which lies not solely in overt, externally observable behavior but also in the internalized arrangement of ideas most of which are predetermined by the discipline. Time is different in history, in physics, in biology, in mathematics, and in English (Henry, 1986, p. 20).

Henry's views concerning the importance of content-specific conceptions of learning and teaching have been well supported in recent years by the work of several teacher educators, such as Buchman (1984).

> Curriculum practices and development in many schools and colleges of education can be interpreted as a flight away from content. Teachers without content are like actors without scripts. Teaching is conditional on the presence of educational content and essential activities of teaching are conditional upon the content knowledge of teachers (pp. 29–30).

The importance of content knowledge in the teaching process has also been the primary focus of study of Shulman and his associates. They have identified general pedagogical knowledge, subject matter knowledge, and pedagogical content knowledge as critical components of the professional knowledge base in teaching (Wilson, Shulman, and Richert, 1987). Pedagogical content knowledge is a relatively new and illuminating construct that refers to the "capacity of a teacher to transform the content knowledge he or she possesses into forms that are pedagogically powerful and yet adaptive to the variations in ability and background presented by the students" (Shulman, 1987, p. 15). Included among the various aspects of pedagogical content knowledge are: (1) the teacher's view of how the discipline should be represented to students; (2) the teacher's understanding of how easy or difficult particular concepts will be for specific groups of students to learn; and (3) the teacher's possession of a variety of examples, metaphors, analogies, and narratives that can be used to make the concepts in the discipline more understandable for students. Wilson, Shulman, and Richert (1987) see the teacher's pedagogical content knowledge as a critical attribute in the process of preparing for, delivering, and reflecting on instruction.

In short, to paraphrase Shulman (1990), when the content to be taught becomes a starting point for the process of inquiry and researchers begin to ask what is good teaching of mathematics or what is good teaching of *Romeo and Juliet,* the answers and related questions seem to be quite different from the answers received when one begins by asking what is good teaching in general. These six beliefs, which characterize the changing mindscape on learning and teaching, call into serious question the portrait that we painted earlier of the supervisor who works one on one with each teacher to document observable behavior and move that behavior into greater alignment with the research on general teaching effectiveness. Indeed, the new mindscape on learning and teaching demands a significantly altered mindscape on supervision.

Implications for Supervision

The changing perspectives on learning and teaching have five important implications:

1. Teachers should be viewed as active constructors of their own knowledge about learning and teaching.

2. Supervisors should be viewed as collaborators in creating knowledge about learning and teaching.

3. The emphasis on data collection during supervision should change from almost total reliance on paper-and-pencil observation instruments to capture the events of a single period of instruction to the use of a variety of data sources to capture a lesson as it unfolds over several periods of instruction.

4. Both general principles and methods of teaching as well as content-specific principles and methods of teaching should be attended to during the supervisory process.

5. Supervision should become more group oriented rather than individually oriented.

Teachers as Active Knowledge Constructors

Just as students must actively construct new knowledge, teachers must be active participants in constructing their own knowledge. The mindscape that has been dominant in supervision has viewed supervision and staff development as vehicles for training teachers to adopt practices and to use knowledge that has been produced by others, principally by researchers on teaching. Just as it is impossible for teachers to pour their knowledge into the heads of students, it is equally impossible for supervisors and staff developers to pour the knowledge and practices recommended by researchers into the heads of teachers. Teachers who choose to adopt new practices are not empty vessels to be filled with someone else's ideas. They are learners who are reeducating themselves to become experts in another mode of teaching (Putnam, 1990). Much of our knowledge about learning remains unused in classrooms not because teachers are unwilling to use it, but because they have not been given the opportunity and the time to work with the concepts and practices in order to relate them to their own knowledge, experience, and contexts—to truly make them their own. Before teachers can use a new model of teaching effectively, they must acquire a deep, personalized understanding of the model. Support for this statement can be derived from the work of Joyce and Showers (1988), which demonstrates that at least thirty to forty hours of study, practice, and feedback are required before teachers gain executive control over complex teaching models. Executive control means that the trainer can use the model well technically, can distinguish between appropriate and inappropriate opportunities for applying the model, and can adapt the model to particular students and contexts.

Perhaps most important, teachers must be looked on as generators of knowledge on learning and teaching, not merely as consumers of research. "What is missing from the knowledge base for teaching, therefore, are the voices of teachers themselves, the questions teachers ask, the way teachers use writing and intentional talk in their work lives, and the interpretive frames teachers use to understand and improve their own classroom practice" (Cochran-Smith and Lytle, 1990. p. 2).

When driven by the new mindscapes on learning and teaching, supervision becomes a vehicle for inquiry and experimentation—aimed at knowledge generation, not simply knowledge adoption. The primary purpose of supervision becomes *the improvement of teaching and learning by helping teachers acquire a deeper understanding of the learning-teaching process.* Knowledge generation can be achieved when supervision becomes a process of action research in which the supervisor and the teacher use classroom learning and teaching activities as a vehicle for testing their own ideas, ideas and practices of colleagues, and findings derived from more formal research studies in terms of their application to the unique educational context in which the teacher and supervisor function.

This view of supervision has been advocated quite powerfully by Schön (1989) and Garman (1986). Garman has taken the view of clinical supervision espoused by Cogan (1973), one of the originators of clinical supervision, and expanded it to be more compatible with current perspectives on learning and teaching. Cogan's model of supervision was grounded in the traditional views of learning, which saw the teacher as the adopter of practices that had been shown as effective through the work of researchers and developers. He did not envision teachers as researchers (Garman, 1986). Garman, on the other hand, points out the necessity for clinical supervisors to engage teachers in the process of self-supervision through reflection and knowledge generation. "At some point in a teacher's career, he/she must become a clinical supervisor of sorts because only; the actors themselves can render the hermeneutic knowledge needed to understand teaching" (Garman, 1990, p. 212). When teachers engage in the process of generating knowledge about their own teaching, they realize important benefits. "Their teaching is transformed in important ways: they become theorists articulating their intentions, testing their assumptions, and finding connections with practice" (Cochran-Smith and Lytle, 1990, p.8).

Supervisors as Collaborators in Creating Knowledge

Just as the teacher's role will change when students are seen as active partners in constructing knowledge, so too the supervisor's role will change when teachers are viewed as constructors of their own knowledge about learning and teaching. From its traditional perspective, supervision is viewed as a process intended to help teachers improve instruction. The supervisor often, intentionally or unintentionally, takes on the role of critic whose task is to judge the degree of congruence between the teacher's classroom behavior and the model of teaching that the teacher is trying to implement or the generic research on teaching.

When the supervisor is viewed as a critic who judges the teacher's performance, supervision tends to concentrate on surface-level issues because the supervisor is denied access by the teacher to the dilemmas, issues, and problems that every teacher experiences and struggles with on an ongoing basis (Blumberg and Jonas, 1987). These dilemmas and problems reach to the very heart of the teaching enterprise and cannot be resolved by simply adding new models to our repertoires of teaching behaviors. They must be confronted head on and resolved through action and reflection in the classroom (Schön, 1983). Supervision should play a central role in understanding and resolving complex, perennial problems such as:

- how to reconcile individual student needs and interests with group needs and interests;

- how to balance the need to preserve student self-esteem with the need to provide students with honest feedback on their performance;

- how to balance student motivation against the need to teach prescribed content that may not match students' current needs or interests; and

- how to maintain a reasonable amount of order while still allowing sufficient flexibility for the intellectual freedom needed to pursue complex topics and issues.

When the supervisor relinquishes the role of critic to assume the role of co-creator of knowledge about learning and teaching, the teacher is more willing to grant the supervisor access to these core issues and dilemmas of teaching because the teacher does not have to fear a critique from the supervisor. Relinquishing the role of critic also benefits the supervisor by removing the awesome burden of serving as judge, jury, and director of the supervisory process. When supervision is viewed as a process for generating knowledge about learning and teaching, data collection is transformed from a mechanism for documenting behavior to a mechanism for collecting information. This information can be used to deepen both teacher's and supervisor's understanding of the consequences of resolving problems, dilemmas, and issues in alternative ways. Conferences are also transformed. In the traditional conference scenario, the supervisor provides a neat, well-documented list of praiseworthy behaviors as well as some suggestion for future improvement. When the supervisor relinquishes the role of critic, conferences become collaborative work sessions in which both teacher and supervisor try to make sense of the almost always messy data that are gathered in the process of relating teacher action to its consequences for learners. Finally, the outcomes of conferences are transformed. In most current practice, both partners sign written narrative critiques, which are filed away to collect dust until next year's observation. When teacher and supervisor become co-creators of knowledge, they produce jointly developed, tentative understandings of the learning-teaching process. These insights can then be tested against the reality of the classroom in future cycles of supervision.

To engage effectively in inquiry-oriented supervision, supervisors need a different type of expertise. They will need a passion for inquiry; commitment to developing an understanding of the process of learning and teaching; respect for teachers as equal partners in the process of trying to understand learning and teaching in the context of the teacher's particular classroom setting; and recognition that both partners contribute essential expertise to the process. They will also need to feel comfortable with the ambiguity and vulnerability of not having prefabricated answers to the problems that are encountered in the process. Supervisors will need to trust themselves, the teacher, and the process enough to believe that they can find reasonable and workable answers to complex questions and problems.

Greater Variety in Data Collection

The emphasis in traditional conceptions of learning on observable behavior, coupled with the emphasis on the teacher as the central actor, has resulted in the use of paper-and-pencil observation instruments as the primary and often sole vehicle for data gathering in supervision. When the supervisor's task is viewed as capturing the observable behavior of one actor (the teacher), paper-and-pencil instruments seem to work reasonably well. However, when learning is viewed as an active process of knowledge construction by the learner, student cognition becomes the critical element in the learning process. Learning is then seen as a collaborative process between teacher and learner, and the task of gathering useful data changes dramatically. Now, the data-gathering task becomes one of simultaneously capturing information about multiple actors which can be used to make inferences about the thinking processes that are occurring in the minds of the actors. This type of data collection requires supplementing paper-and-pencil instruments with a wide range of data-gathering techniques including audiotapes, videotapes, student products (essays, projects, tests), student interviews, and written student feedback regarding classroom events.

The use of multiple sources of data will bring about another important change in the expertise required of those who function as supervisors. The supervisor will need to become an expert in helping the teacher match various types of data collection strategies to the questions that are being addressed in the supervisory process and in helping the teacher interpret and reflect on the data that have been gathered. This change in the focus of data collection techniques will parallel closely the changes that have taken place in educational research techniques over the past decade. Just as the paper-and-pencil instruments used in the process-product research on teaching have been augmented by qualitative data collection strategies, so too observation and data collection in supervision can be expanded to include many more data sources. Data alone, however, are never sufficient. They never tell the full story. Only human judgment, in this case the collaborative judgment of teacher and supervisor, can give meaning to the richness of the learning-teaching process. Human judgment functions much more effectively in capturing that richness when it is augmented by a wide variety of data sources.

Garman (1990) points out an additional factor that comes into play when we view the goal of data collection as capturing student and teacher thinking: the development of thinking over time. Data collection currently is almost always accomplished by the observation of a single period of instruction.

> [A] lesson generally means an episodic event taken out of context within a larger unit of study. It is time to consider the unfolding lesson as a major concept in clinical supervision. We must find ways to capture how a teacher unfolds the content of a particular unit of study and how students, over time, encounter the content (Garman, 1990, p. 212).

By collecting data over longer periods of instruction, we would be likely to obtain a much more complete picture of both teacher and student thinking. We would also capture a much richer portrait of the teacher's view of how the discipline should be represented for students. Although it might at first seem that collecting data over several periods of instruction requires additional time for observation by the supervisor, this is not necessarily the case. When the teacher becomes a collaborator in the process, and multiple data collection techniques are used (e.g., videotapes, student homework, student tests), the supervisor need not be present for every period of instruction during which data are gathered. The teacher can take primary responsibility for much of the data collection and then meet with the supervisor to jointly interpret and discuss the meaning of the data.

Greater Balance Between General Concerns and Content-Specific Issues and Questions

Given the renewed emphasis and research on content-specific learning and teaching, the focus of supervision should shift from total emphasis on general concerns to the inclusion of content-specific issues and questions. This does not mean that we should exclude general behaviors. To do so would clearly be a mistake because process-product research has been successful in identifying some behaviors that seem to transfer across content (Gage and Needels, 1989). However, as Shulman (1987) has pointed out, excluding content-specific strategies from the supervisory process has also been a mistake. We need to balance content-specific issues and general issues.

On the surface at least, this need to expand the focus of supervision poses a dilemma for many schools. Principals, who supervise teachers in many different content areas, carry out much of the supervision that takes place in schools. The question is whether a generalist can be an effective supervisor when the supervisory process focuses not only on general concerns but also on content-specific strategies and methods. Given the new supervisory mindscape, we believe it is possible.

If the supervisor is viewed as a collaborator whose primary task is to help teachers reflect on and learn about their own teaching practices through the collection and interpretation of multiple sources of data, and the teacher who has content expertise is allowed to direct the process, it seems reasonable to think that content-specific issues could be addressed through supervision. In addition, if supervision is viewed as a function—not merely a role—to which many people in a school can contribute (Alfonso and Goldsberry, 1982), it would also be possible to use a process of group supervision, peer coaching, or colleague consultation to help address content-specific issues, provided the peers have the appropriate preparation and skills.

Whatever personnel are used to carry out the process, the scope of supervision needs to be expanded to include questions such as these: What content should be taught to this group of students? Are the content and the instructional approaches being used compatible? What beliefs about the content and its general nature are being conveyed to students by the teacher's long-term approach to the subject matter? Are students acquiring the thinking and learning strategies that are most important for long-term success in the discipline?

Emphasis on Group Supervision

Just as students seem to benefit when they are placed in groups to cooperate with each other in the learning process, teachers seem to benefit when they are allowed to work together in groups to help each other learn about and refine the process of teaching (Little, 1982). Teachers learn by watching each other teach. In addition, the new roles they take on and the perspectives they gain promote higher levels of thinking and cognitive development (Sprinthall and Thies-Sprinthall, 1983). This benefits students because teachers who have reached higher cognitive-developmental levels tend to be more flexible and better able to meet individual student needs (Hunt and Joyce, 1967). Collaborative practices have been endorsed and employed in staff development circles for several years; however, supervisory practice, which also aims at professional development, typically continues to occur on a one-to-one basis between supervisor and teacher.

We concur with Fullan (1990), who pointed out the necessity of linking collaboration to norms of continuous improvement:

> There is nothing particularly virtuous about collaboration per se. It can serve to block change or put students down as well as to elevate learning. Thus, collegiality must be linked to norms of continuous improvement and experimentation in which teachers are constantly seeking and assessing potentially better practices inside and outside their own school (p. 15).

Similarly, group supervision must be viewed as an activity whose primary aim is learning about and improving teaching. Teachers are sometimes uncomfortable when they are asked to confront tough questions about their own teaching. Collaboration and mutual support from colleagues can be vehicles for enabling teachers to risk facing these tough questions. However, there is a danger that collaboration can be wrongly viewed as meaning to support one another without rocking the boat or causing any discomfort. When this happens, collaboration can degenerate into a mechanism for skirting tough questions through unwarranted assurances that things are just fine. To avoid this degeneration, all participants must understand that learning about the instructional process and improving student learning are the primary goals of group supervision. Collaboration is a means to an end, not an end in itself. It is a mechanism for providing support as teachers engage in the sometimes disquieting, uncomfortable process of learning.

Given the research on cooperative learning and teacher collegiality, we hypothesize that if supervision were carried out as a group process in which the supervisors and teachers were interdependent in achieving group and individual goals, the process of supervision would become more effective in helping teachers learn about and improve their teaching. In addition, enabling those teachers who may be less committed to growth to work together in groups with colleagues who are more committed to the process may be an effective strategy for creating shared norms that are supportive of the supervisory process. In discussing the concept of collaborative cultures, Hargreaves and Dawe (1989) eloquently describe what supervision might become when it is viewed as a cooperative group process. "It is a tool of teacher empowerment and professional enhancement, bringing colleagues and their expertise together to generate critical yet also practically-grounded reflection on what they do as a basis for more skilled action" (p. 7).

◆ ◆ ◆

What we have labeled "the changing mindscape on learning and teaching" demands a new mindscape on supervision, a mindscape grounded in the following principles and beliefs:

1. The primary purpose of supervision is to provide a mechanism for teachers and supervisors to increase their understanding of the learning-teaching process through collaborative inquiry with other professionals.

2. Teachers should not be viewed only as consumers of research, but as generators of knowledge about learning and teaching.

3. Supervisors must see themselves not as critics of teaching performance, but rather as collaborators with teachers in attempting to understand the problems, issues, and dilemmas that are inherent in the process of learning and teaching.

4. Acquiring an understanding of the learning-teaching process demands the collection of many types of data, over extended periods of time.

5. The focus for supervision needs to be expanded to include content-specific as well as general issues and questions.

6. Supervision should focus not only on individual teachers but also on groups of teachers who are engaged in ongoing inquiry concerning common problems, issues, and questions.

These principles and beliefs are not completely new. They closely parallel the principles of clinical supervision as endorsed by Cogan (1973) and Goldhammer (1969). Unfortunately, these principles have not been widely adopted. We believe that the changing perspectives on learning and teaching provide a powerful impetus for putting these principles of supervision into practice. When these concepts begin to touch the mainstream of supervisory practice, supervision is much more likely to have a positive impact on teacher thinking, teacher behavior, and student learning.

References

Alfonso, R.J., and L. Goldsberry. (1982). "Colleagueship in Supervision." In *Supervision of Teaching,* edited by T.J. Sergiovanni. Alexandria, Va.: ASCD.

Anderson, L.M. (1989). "Classroom Instruction." In *Knowledge Base for the Beginning Teacher,* edited by M.C. Reynolds. New York: Pergamon Press and the American Association of Colleges of Teacher Education.

Blumberg, A., and R.D. Jonas. (1987). "Permitting Access: The Teacher's Control Over Supervision." *Educational Leadership* 44, 8: 12–16.

Brandt, R. (1988–89). "On Learning Research: A Conversation with Lauren Resnick." *Educational Leadership* 46, 4: 12–16.

Brandt, R. (1989–90). "On Cooperative Learning: A Conversation with Spencer Kagan." *Educational Leadership* 47, 4: 8–11.

Bransford, J.D., and N.J. Vye. (1989). "A Perspective on Cognitive Research and Its Implications for Instruction." In *Toward the Thinking Curriculum: Current Cognitive Research,* edited by L.B. Resnick and L.E. Klopfer. Alexandria, Va.: ASCD.

Brooks, J.G. (1990). "Teachers and Students: Constructivists Forging New Connections." *Educational Leadership* 47, 5: 68–71.

Buchman, M. (1984). "The Priority of Knowledge and Understanding in Teaching." In *Advances in Teacher Education* Vol. 1, edited by L.G. Katz and J.D. Raths. Norwood, NJ.: Ablex.

Cochran-Smith, M., and S.L. Lytle. (1990). "Research on Teaching and Teacher Research: Issues That Divide." *Educational Researcher* 19, 2: 2–11.

Cogan, M. (1973). *Clinical Supervision.* Boston: Houghton-Mifflin.

Dewey, J. (1902). *The Child and the Curriculum.* Chicago: University of Chicago Press.

Fullan, M. (1990). "Staff Development, Innovation, and Institutional Development." In *Changing School Culture Through Staff Development. The 1990 ASCD Yearbook,* edited by B. Joyce. Alexandria, Va.: ASCD.

Gage, N.L., and M.C. Needels. (1989). "Process-Product Research on Teaching: A Review of Criticisms." *Elementary School Journal* 89, 3: 253–300.

Garman, N.B. (1986). "Reflection: The Heart of Clinical Supervision: A Modern Rationale for Professional Practice." *Journal of Curriculum and Supervision* 2, 1: 1–24.

Garman, N.B. (1990). "Theories Embedded in the Events of Clinical Supervision: A Hermeneutic Approach." *Journal of Curriculum and Supervision* 5, 3: 201–213.

Goldhammer, R. (1969). *Clinical Supervision: Special Methods for the Supervision of Teachers.* New York: Holt, Rinehart, and Winston.

Hargreaves, A., and R. Dawe. (1989). "Coaching as Unreflective Practice." Paper presented at the Annual Meeting of the American Educational Research Association, San Francisco.

Henry, G. (1986). "What Is the Nature of English Education?" *English Education* 18, 1: 4–41.

Hunt, D.E., and B.R. Joyce. (1967). "Teacher Trainee Personality and Initial Teaching Style." *American Educational Research Journal* 4: 253–259.

Joyce, B., and B. Showers. (1988). *Student Achievement Through Staff Development.* New York: Longman.

Lampert, M. (1990). "When the Problem is Not the Question and the Solution Is Not the Answer: Mathematical Knowing and Teaching." *American Education Research Journal* 27, 1: 29–63.

Little, J. (1982). "Norms of Collegiality and Experimentation: Workplace Conditions of School Success." *American Educational Research Journal* 5, 19: 325–340.

Perkins, D.N., and G. Salomon. (1989). "Are Cognitive Skills Context-Bound?" *Educational Researcher* 8, 1: 16–25.

Putnam, R. (1990). "Recipes and Reflective Learning: 'What Would Prevent You from Saying It That Way?'" Paper presented at the Annual Meeting of the American Educational Research Association, Boston.

Resnick, L.B., and L.E. Klopfer. (1989). *Toward the Thinking Curriculum: Current Cognitive Research.* Alexandria, VA: ASCD.

Schlechty, P.C. (1990). *Schools for the 21st Century.* San Francisco: Jossey-Bass.

Schön, D.A. (1983). *The Reflective Practitioner.* San Francisco: Jossey-Bass.

Schön, D.A. (1989). "Coaching Reflective Teaching." In *Reflection in Teacher Education,* edited by P.P. Grimmet and G.P. Erickson. New York: Teachers College Press.

Sergiovanni, T.J. (1985). "Landscapes, Mindscapes, and Reflective Practice in Supervision." *Journal of Curriculum and Supervision* 1, 1: 5–17.

Shulman, L.S. (1987). "Knowledge and Teaching: Foundations of the New Reform." *Harvard Educational Review* 57: 1–22.

Shulman, L.S. (1990). "Transformation of Content Knowledge." Paper presented at the Annual Meeting of the American Educational Research Association, Boston.

Slavin, R.E. (1989–90). "Research on Cooperative Learning: Consensus and Controversy." *Educational Leadership* 47, 4: 52–54.

Sprinthall, N.A., and L. Thies-Sprinthall. (1983). "The Teacher as Adult Learner: A Cognitive Developmental View." In *Staff Development. 82nd Yearbook of the National Society for the Study of Education,* edited by G.A. Griffin. Chicago: University of Chicago Press.

Wilson, S.M., L.S. Shulman, and A.E. Richert. (1987). "150 Different Ways of Knowing: Representations of Knowledge in Teaching." In *Exploring Teachers' Thinking,* edited by J. Calderhead. London: Cassel.

SESSION 2

Building Intellectual Communities in Mathematics Classrooms

In this session you will begin to develop a new eye for the mathematics classroom. First, you will reflect on the intellectual community in the classroom you are observing on videotape and the qualities that exemplify such a community. You will also think about how the interactions between teachers and students cultivate and sustain the intellectual community. We look at the intellectual community first since many administrators already attend to that when they do classroom observations.

However, we would like to suggest that you look beyond the procedural and interactive features and think about the interrelationship between these features and the mathematical content being explored. For some of you, this focus will represent a shift in what you pay attention to in classrooms. You may readily recognize features of a supportive community, such as the level of trust between teacher and students or the level of comfort students feel taking risks, but you may not be used to looking at how the classroom setting supports an environment of inquiry into important mathematical ideas. This environment of inquiry is essential in a standards-based classroom where both the quality and depth of ideas explored are important and, as observers, you need to be able to discern not only whether the classroom is a supportive community but whether the community fosters a fertile environment for mathematical inquiry.

The activities in this session are designed to encourage you to look beyond the surface features of the classroom as you describe the intellectual community in the classroom and to consider how this community supports the exploration of substantive mathematical ideas.

Learning and Teaching Mathematics

In preparation for Session 2, read *Learning Mathematics While Teaching,* by Russell, Schifter, Bastable, Yaffee, Lester, and Cohen. The authors look at three elementary grade teachers who "learn" mathematics in the context of their own teaching. As you read the article, consider the following questions:

1. What do the authors mean by "learning" mathematics while teaching?

2. What knowledge do teachers need to have to engage in the kind of learning of mathematics suggested by the authors?

3. Consider one of the three cases presented by the authors. How does being open to learning mathematics as this teacher was support the cultivation of a mathematical intellectual community in the classroom?

4. What seems appropriate and/or challenging about being such a learner, given what you know about the realities of teaching in your own school?

Learning Mathematics While Teaching*

Susan Jo Russell, Deborah Schifter, Virginia Bastable, Lisa Yaffee, Jill B. Lester, and Sophia Cohen

This paper examines cases of elementary grade teachers learning mathematics in the context of their own teaching, as they explore mathematics content they are using with their students, consider student strategies and representations that are new to them, and try to understand how students are thinking about complex mathematical ideas. We consider what teachers must already understand in order to do this and discuss implications for teacher education.

It is widely recognized that in order to teach mathematics for understanding, teachers, themselves, whose mathematical experiences have been limited to traditional instruction, need to understand the content more deeply than most currently do (Ball, 1991; Cohen *et al.*, 1990; Schifter, 1993). Although some programs have provided opportunities for teachers to explore significant mathematics content (Lappan & Even, 1989; Russell & Corwin, 1993; Schifter & Fosnot, 1993; Simon & Schifter, 1991), it is unclear just what teachers need to learn to be able to support their students' constructions of rich mathematical concepts. In fact, it appears that the new mathematical understandings teachers must develop and the teaching situations they must negotiate are too varied, complex, and context-dependent to be anticipated in one or even several courses. Thus, teachers must become learners in their own classrooms (Ball, in press; Featherstone *et al.*, 1993; Heaton, in press).[1]

In Teaching to the Big Ideas, a joint project of EDC, TERC, and SummerMath for Teachers, we are exploring the development of teachers' mathematical understandings and their effect on instruction (Schifter *et al.*, in preparation). Data include classroom field notes, audiotaped interviews, and papers and journals written by the teachers which include reflections on episodes in their classrooms. The project began in the summer of 1993. Drawing from our work in Teaching to the Big Ideas during the 1993–1994 and 1994–1995 school years, this paper examines cases of teachers learning mathematics in the context of their own teaching and considers what teachers must already understand in order to do this.

---

*from *Inquiry and the Development of Teaching: Issues in the Transformation of Mathematics Teaching* edited by Barbara Scott Nelson. "Learning Mathematics While Teaching" by Susan Jo Russell, Deborah Schifter, Virginia Bastable, Lisa Yaffee, Jill B. Lester, and Sophia Cohen. Newton, MA Center for the Development of Teaching, Education Development Center, Inc. Copyright ©1995. Reprinted by permission. All rights reserved.

What Do We Mean by Learning Mathematics?

When we first asked teachers to identify episodes in their classrooms during which they learned mathematics, they were stymied. To many of them, "to learn" seemed to mean either the acquisition of completely new knowledge about previously unfamiliar mathematics or, perhaps, an "Aha!" experience in which an idea is apprehended for the first time. We propose to extend our definition of learning mathematics to include a more ongoing and gradual process in which understanding of familiar content is deepened as one makes new connections and distinctions. A new representation of mathematical relationships may illuminate an aspect that was previously invisible even though that relationship was already "known" or "understood." An unfamiliar problem or context may highlight a mathematical idea in a new way, making one's thinking more problematic and causing one to think more explicitly about what was implicitly known.

It is in the context of this view of learning mathematics—a gradual building and deepening—that we offer glimpses of teachers engaged in learning mathematics in the course of teaching. As is the case when we observe student learning, we see only a slice and often do not know the whole story. However, we believe these episodes, taken together, provide evidence of what we mean by teachers becoming learners of mathematics in their classrooms. This paper considers episodes that illustrate teachers engaged in learning mathematics in three contexts: 1) exploring a mathematical problem or question embedded in content they are teaching; 2) thinking through students' representations and strategies; and 3) looking underneath students' confusions or excitement to consider mathematical structures.

1. Exploring Mathematics Content

One of the ways we see teachers learning mathematics is by engaging directly in the mathematical content they are teaching their students. Before they begin a unit, teachers might explore a mathematics topic by reviewing several resource books, solving some problems, or discussing the issues of content with colleagues. But sometimes it is only after plunging into work with students that teachers identify mathematical issues that they want to explore further for themselves.

Meg gave her second graders a word problem that she had not investigated ahead of time: What are some possible combinations of 12 marbles if each marble can be one of three colors? In other words, each combination of 12 marbles could include red and/or green and/or blue marbles. So, legitimate combinations might be 12 red marbles; 6 red marbles and 6 blue marbles; 3 green marbles, 8 green marbles, and 1 blue marble. Once she observed her students working on it, Meg realized the problem was more complex than it had first seemed and decided she needed to understand it better. At home, she worked on the problem with her husband. They began with 1 marble to find the number of combinations possible with 3 colors (3 combinations & 1 blue, 1 green, 1 red) then moved on to consider 2 marbles (6 combinations: 2 blue, 2 green, 2 red, 1 blue and 1 green, 1 blue and 1 red, 1 green and 1 red), 3 marbles (10 combinations), and so forth up to 12 marbles. She noticed that the differences increased by 1 (with 2 marbles, there are 3 more combinations than with 1 marble; with 3 marbles, there are 4 more combinations than with 2 marbles; etc.), and wondered what would happen with four colors. Meg pursued this problem for 12 pages in her journal, and then commented:

> I have a very hazy, in-and-out picture of why I can make that number of combos with four colors of marbles and why it grows the way it does as you add a marble—because by adding 1, you're adding all these other possibilities. I could NEVER explain it at this point. AND, I don't get the relationship between what happens with two colors, three colors, five colors beyond that it grows REALLY FAST . . .

When Meg originally presented this problem to her class, she had not thought through the mathematical ideas embedded in the problem. She knew that it was a type of problem that calls for generating combinations systematically, but she did not know enough about the mathematics that arises in this problem to develop a mathematical goal for her students. As she worked on the problem for herself, she discovered a number of things, including the large number of combinations that can be generated, the need for systematicity in order to generate all the combinations, and the pattern of increases as marbles are added. She began to think through why the pattern of increases works as it does—that adding one marble adds "all these other possibilities"—and noticed that increasing the number of either marbles or colors leads to a rapid increase in the number of possible combinations.

Now that she had a deeper understanding of the complexity of the problem, Meg turned her attention back to the thinking of her students and the issue of how to proceed in her classroom:

> What are the implications for asking seven year-olds to work on a problem with 91 answers? . . . The kids who approached it randomly thought about 12-ness, and adding numbers, and three parts—and some kids went further. But now I want to ask them something more manageable to see if some "finish" and know they're done.

She chose to pose the following problem to her class: "There are 23 cupcakes for our good-bye show. Some have white frosting. The rest have green frosting. How many of each color frosting might there be?" This problem presents only two, instead of three, color possibilities, making it more possible for students at this age level to work on generation, comparison, and organization of combinations and to decide whether or not they have found all possible combinations. In working through the mathematics of the original problem, we surmise that Meg began to sort through the mathematical ideas that arise in a combinatorics problem and rejected some of these ideas as inappropriate for her second graders. She knew, for example, that for the marbles problem, her students would not be able to organize the many possibilities they generated and so would not be able to explore growth patterns or recognize whether or not they had all of the possible combinations. Because of her own thinking about why growth occurred in the way that it did, she realized that of the two questions—Do I have all the combinations? and Can I predict the number of combinations?—The first was the most accessible for her students. Meg then chose a mathematical goal that focused on systematicity rather than on growth patterns. The problem has actually been altered dramatically by moving from three colors to two colors. Now the problem focuses on the number of ways of breaking one quantity into two parts. While it involves work on combinations, it begins at a point that is very close to the work these second graders are doing as they become fluent in breaking numbers into manageable parts to solve addition and subtraction problems. By reducing the problem to two colors, Meg suspected that at least some of her second graders would be able to find a systematic way to keep track of the possible combinations of white and green frosting for 23 cupcakes and would be able to begin to look at patterns in their combinations (1 + 22, 2 + 21, 3 + 20, . . .). By learning more about the mathematics of combinatorics problems she was able to devise a problem with an appropriate mathematical goal for her students.

This clearer view of the mathematics content also leads, we suspect, to the posing of more focused questions to students as they work. Choosing problems and linking them with clear mathematical goals is one of the key tasks of a teacher of mathematics. Learning more about the mathematical ideas embedded in a problem or class of problems supports the teacher in making choices about creating, choosing, or modifying problems for the students and in interactions with students' developing ideas.

2. Thinking through Students' Representations and Strategies

Another way in which we observe teachers learning mathematics is as they engage in thinking through students' approaches to solving problems. Learning takes place as teachers are confronted by student strategies or representations that are different from their own. In assessing the reasoning of students' responses, especially when they are unfamiliar and unexpected, teachers think through the mathematics again for themselves, seeing new aspects of familiar content, expanding their own understandings.

Denise, a second-grade teacher, expanded her view of the process of subtraction when she observed one of her students develop an algorithm that, she later wrote, "I had never thought of, or even imagined before." Denise was working with Ivan and his partner after he had made the familiar subtract-the-smaller-from-the-larger error in the problem $52 - 28$, getting an answer of 36. Brandon said to Ivan, "But you can't take the 2 away from the 8; you have to take the 8 away from the 20." As Ivan began to rethink his solution and Brandon's comment, Denise expected to work out something about "borrowing a 10 from the 50 and adding it to the 2 to make 12, which would match her own representation of the problem. But, instead, Ivan invented a different method, "You take the 20 away from the 50 and get 30. Then you take the 8 away from the 2 which is minus 6. Then you take the minus 6 away from 30 and you get 24." Denise reported that she had to ask Ivan to repeat his solution several times before she understood his method. She commented in her journal:

> I've since read Connie Kamii's book in which she describes several common methods 2nd graders use to subtract in this situation and [Ivan's] method was one of them. It still feels new enough to me that I have to think it through each time. It is definitely a case of my learning some mathematics from my students.

Throughout their study of addition and subtraction, Denise encouraged students in her classroom to develop their own computation strategies. As part of this work, she emphasized pulling apart numbers in a variety of ways to make the addition or subtraction process more manageable. For example, methods in her classroom for this problem included strategies such as these:

OR $52 - 20 = 32; 32 - 2 = 30; 30 - 6 = 24$
$50 - 20 = 30; 30 - 8 = 22; 22 + 2 = 24$

So Denise was already comfortable with methods that involved breaking numbers into parts in a variety of ways and recombining them to solve the problem. What makes the problem $52 - 28$ different is the embedded subproblem $2 - 8$. Many of the strategies invented by Denise's students transform a problem like this one into a set of subproblems in a way that eliminates the larger-from-smaller subtraction (e.g., $2 - 8$). For example, in the methods above, the 20 (from the 28) is subtracted first, then the 8 is subtracted from the result. In the first method, the 8 is broken into 2 and 6 in order to subtract easily from the 32, while in the second method, the 2 from the 52 is eliminated from the problem, then added back to the result in the final step. However, it had not occurred to Denise in her own thinking that these numbers could be pulled apart in such a way that the larger-from-smaller sub-problem would be used as it is, leading to a negative result that would then be recombined with other parts of the problem. It also may not have occurred to her that one of her second graders would feel comfortable going "below zero" as part of the subtraction process. In coming to understand Ivan's method, Denise broke her own "below zero" barrier in thinking about what is allowed in solving a subtraction problem.

As another example, consider Ellen, who had been working on multiplication and division with her third-grade class: Multiplication was presented as repeated groups, while division was modeled as dealing out. These were models of the operations with which Ellen herself was comfortable. During one class, Ellen asked Kevon to illustrate 7×3 on the board. Kevon began by drawing seven circles. He then drew one mark inside each circle, paused, and drew a second mark in each circle. Ellen recognized these actions as what she and her students usually did for a division problem. She was about to have Kevon sit down to give someone else a chance, but she caught herself hesitating as she studied the board. She asked the class, "What is Kevon doing? It looks like what we usually do for division but let's wait and see what he comes up with." Kevon then added a third line to each of his seven circles and wrote "21" to the right of the equals sign in his equation. Ellen asked Kevon how he came up with 21 and Kevon explained, "I put 1 of each, 2 of each, 3 of each." Ellen asked, "What does that mean?" Kevon responded, "I got 7 of them with 3 in each." Ellen thought about this for a long moment before answering. At this point in the interaction, she was challenged by Kevon's representation to expand her own model of multiplication. She said, almost to herself, "Seven groups of what? Seven groups of three. Should we consider what he did right?"

Once she and the children decided Kevon's representation fit the problem, Ellen asked the class if anyone had still other diagrams. The staff observer noted that Ellen's acceptance of alternative representations was unusual for her. In response, students offered a greater variety of solution methods than had previously been the case during the group's study of multiplication and division. We surmise that her recognition of Kevon's unfamiliar representation—beginning with 7 groups and then dealing out marks until each group had 3, rather than the iterating of threes she and her class had been using acted as a catalyst to her invitation to students to show their own ways of thinking about multiplication.

In both of these episodes, teachers took time to understand students' approaches. In both cases, they not only learned about their students' thinking, but they actually expanded their own views of ways to model a whole number operation.

3. Delving Underneath Students' Reasoning: Looking at Mathematical Structure

In the third kind of example, teachers' reflections on what is problematic in students' reasoning leads them to rethink their own understanding of mathematical structures. By probing underneath students' confusions about mathematical ideas, they confront new mathematics themselves as they ask these questions: Why would a student think that? What is right about the solution from the student's point of view? What is difficult to understand about the mathematics here and why might that be the case? Often this process involves making explicit and reexamining what they have implicitly known.

For example, students in Sylvia's second-grade classroom were adding two-digit numbers, using base ten blocks to model "carrying." Most students easily undertook this task, following the conventions established in the class for using the blocks and recording the results. However, when some students were asked by a staff observer to determine without trading what quantity was represented by 4 tens sticks and 15 small cubes, they had difficulty counting this quantity. One student counted 10, 20, 30, 40 for the tens sticks, then continued counting the small cubes by tens, and was perfectly satisfied with her result of 190. Sylvia was quite surprised when she observed students counting ones as tens. She had assumed from their competent use of the base ten blocks to solve addition problems that they understood how to decompose two-digit numbers into tens and ones and how to recombine them. She was still thinking about this some months later when she described how various students in her classroom were solving two-digit addition problems, some breaking apart numbers into tens and ones, others counting on only by ones. She commented: "What goes on in someone's brain to make sense of tens and ones? Does a person need to construct a system of ones and a system of tens that sort of 'fits' like an overlay on the system of ones? How does this happen?"

The next year Sylvia again introduced rods trading but also used many word problems throughout the year. She did not insist that students use the rods to solve these problems but encouraged them to develop and discuss many strategies based on their own understanding of the number relationships. She wrote:

> Rods trading had no discernible impact on how the children thought about addition and subtraction problems. Some children, who could answer questions about how many tens, how many ones, where are the tens and ones, and so on, would still count on by ones when solving double-digit addition problems. [Other children] . . . added the way they always had: tens first, and then the ones.

Over two years, as Sylvia closely watched her students working with ones and tens, she began to refine and deepen her own ideas about the structure of the base ten system. Many of her students could easily learn how to manipulate the base ten blocks to come up with the correct solution to an addition problem. However, when operating in an addition or subtraction situation without the blocks, they still did not flexibly use what they seemed to "know" about tens and ones. As Sylvia pondered more deeply what is involved in constructing and simultaneously manipulating more than one unit (e.g., ones and tens), she began to consider how the conception of a unit does not inhere in a particular physical representation (such as base ten blocks) but in a complex mental model in which a larger unit (such as "ten") constructed out of smaller units (such as "ones") can exist simultaneously as a one and a collection of ones. It can be manipulated as a large unit that can be combined with, compared to, or separated from a collection of the same large units as if they were ones. Yet this "one" can also be decomposed into the smaller units of which it is composed, should the need arise.

Certainly, operating with ones and tens was not new to Sylvia. However, in watching her students, Sylvia began to think more deeply about the complexity of coordinating multiple units (in this case, tens and ones). Once the idea of coordinating multiple units—which had been obscured for Sylvia by her students' apparent competence in manipulating base ten blocks to solve computation problems—became explicit for her, she could see this same idea underlying many aspects of her students' mathematical activity in skip counting which she did frequently with her class. Students say, for example, the numbers 2, 4, 6 to represent two, four, six objects while each number also represents one count (one unit). As she moved through the second-grade curriculum, Sylvia found that this now visible idea of multiple units comes up in multiplication and division as well as in problems related to time, money, and measuring. Sylvia's understanding of all these topics has been enriched by the new clarity with which she sees the common underlying idea of multiple units.

What Enables Teachers to Learn Mathematics While Teaching?

As we collect and analyze episodes such as those in this paper, we are beginning to identify some of the elements that appear to be necessary for teachers to learn mathematics while teaching. First of all, this kind of exploration of mathematics content requires teachers to see themselves as adult learners of mathematics and to see their own classrooms as contexts in which they learn. The teachers described here had participated in at least one year of experiences that emphasized adult mathematics learning, and they now share some assumptions: 1) learning about mathematics occurs when one is immersed in problem solving; 2) an important aspect of mathematical thinking is identifying, describing, and testing patterns and relationships; and 3) understanding the mathematics better has implications for their pedagogical decisions. Rather than expecting to acquire discrete bits of information, these teachers assume that the way to learn mathematics is to do mathematics.

In the first category of learning mathematics while teaching, teachers explore the mathematics content in which they engage their students. To do so, they must be curious about mathematics, know that they can pose their own mathematical questions, and assume that they can pursue those questions themselves.

In the second group of episodes, teachers expand their understanding of mathematical ideas by paying attention to student strategies and representations. This requires teachers to develop a classroom culture in which multiple strategies and unexpected responses are the norm, where the point is not to find the solution prescribed by teacher or textbook, but to reason cogently about mathematical relationships. In order to analyze student thinking, teachers must learn how to follow a mathematical argument and assess its validity. As they consider student strategies and representations that are different from their own, teachers become aware of new aspects of mathematical relationships. Revisiting familiar mathematical ideas in this way can lead to a deepening appreciation for their complexity.

In the third category of episodes, teachers delve beneath students' efforts at understanding to confront the underlying mathematical structures with which their students are grappling. By carefully listening to and observing their students, they begin to identify and describe the complexity of elementary mathematics— what mathematical ideas are central to student understanding and why these are hard for students. When teachers reflect on student learning in this way, they are doing more than understanding student thinking better; they are actually recognizing and articulating significant mathematical ideas and developing a deeper understanding of these ideas for themselves.

Do teachers know that they are learning mathematics as they teach? In some of the episodes recounted here, teachers seem to be aware that they are engaged in mathematics learning; in others, perhaps only the authors view the event in this way. Often, we have seen that the simplicity of the formulations brought about in such new learning situations leads teachers to discount or disparage their new knowledge: "Oh, I was so stupid not to see that," or "Of course, I'm sure it was obvious to everyone else, but . . ." On the other side of learning, the simple and elegant might appear trivial—not worth mentioning. Beginning to notice and appreciate the profundity of our own insights as adults is connected to appreciating the profundity of children's insights.

Learning in the context of one's own teaching is not simply a remedial measure. Rather, it is a component of the pedagogy itself and already requires of teachers a fairly sophisticated understanding of the discipline. In order for teachers to take advantage of the opportunities that present themselves in their own classrooms, they need an orientation to what such learning might be like. One of the objectives of teacher education programs can be to prepare teachers for the ways in which they can take advantage of their teaching as a site for their own ongoing learning of mathematics. Making explicit and validating new learning of this kind may be a critical function of teacher education.

Acknowledgments

This work was supported by the National Science Foundation under Grant no. ESI-9254393. Any opinions, findings, conclusions, or recommendations expressed in this paper are those of the authors and do not necessarily reflect the views of the National Science Foundation.

Notes

[1] The classroom is also an important context for learning about teaching and about student thinking. While we in no way want to diminish the importance of these two components, in this paper we highlight how teachers learn *mathematics content* in their own classrooms.

References

Ball, D. (1991). Research on teaching mathematics: Making subject matter knowledge part of the equation. In J. Brophy (Ed.), *Advances in Research on Teaching* (Vol. II.) (1–48). Greenwich, CT: JAI Press.

Ball, D. (in press). Connecting to mathematics as part of learning to teach. In D. Schifter (Ed.), *What's Happening in Math Class? Volume 2: Reconstructing Professional Identities.* New York: Teachers College Press.

Cohen, D.K., Peterson, P.L., Wilson, S., Ball, D., Putnam, R., Prawat, R., Heaton, R., Remillard, J., & Wiemers, N. (1990). *Effects of state-level reform of elementary school mathematics curriculum on classroom practice* (Research Report 90–14). East Lansing MI: The National Center for Research on Teacher Education and The Center for the Learning and Teaching of Elementary Subjects, College of Education, Michigan State University.

Featherstone, H., Smith, S.P., Beasley, K., Corbin, D., & Shank, C. (1993). *Expanding the Equation: Learning Mathematics through Teaching in New Ways.* Paper presented to the American Educational Research Association, Atlanta, GA.

Heaton, R. (in press). Learning while doing: Understanding early efforts to create new practices of teaching and learning mathematics for understanding. In D. Schifter (Ed.), *What's Happening in Math Class? Volume 2: Reconstructing Professional Identities.* New York: Teachers College Press.

Lappan, G., & Even, R. (1989). *Learning to teach: Constructing meaningful understanding of mathematical content* (Craft Paper 89-3). East Lansing MI: National Center for Research on Teacher Learning, Michigan State University.

Russell, S. J., & Corwin, R. B. (1993). Talking mathematics: "Going slow" and "letting go." *Phi Delta Kappan,* 74:(7), 12.

Schifter, D. (1993). Mathematics process as mathematics content: A course for teachers. *Journal of Mathematical Behavior,* 12:(3), 271–283.

Schifter, D., & Fosnot, C.T. (1993). *Reconstructing Mathematics Education: Stories of Teachers Meeting the Challenge of Reform.* New York: Teachers College Press.

Schifter, D., Russell, S. J., & Bastable, V. (in preparation). Teaching to the big ideas. In M. Solomon (Ed.), *Reinventing the Classroom.*

Simon, M.A., & Schifter, D. (1991). Toward a constructivist perspective: An intervention study of mathematics teacher development. *Educational Studies in Mathematics,* 22, 309–331.

SESSION 3

Linking Intellectual Community with Mathematical Inquiry

In this session, you will continue to look at the aspects of intellectual community that are important when engaging in mathematical inquiry. In addition, you will think about topics of discussion at the post-observation conference based on your observations. This last activity introduces the second strand of this course, *Talking with Teachers about Mathematics, Learning, and Teaching,* and the notion of co-inquiry that underlies it.

As was discussed in the last session, you may be accustomed to looking at process features of the classroom, such as the teacher involving students in discussion or students working or talking with each other, many of which are also characteristic of the standards-based classrooms. You may not, however, be used to looking for a clear mathematical intent to these activities to see if they connect to a broader mathematical agenda or at the intellectual community to see if students are engaged fully in the mathematical ideas embedded in the activities.

In this session, you will engage in a mathematics investigation of algebraic concepts that underlie an elementary math problem. By investigating the same problem as the students in the videotaped classroom episode, you can understand the kind of thinking that helps students establish the mathematical foundation they will need when they begin their study of formal algebra.

Observing in Classrooms: Building an Intellectual Community

In the next session, we will be continuing our exploration of a mathematical intellectual community. We will begin to examine what it means to think mathematically and will also look more carefully at how students participate in intellectual communities. In preparation for this session, please do the following:

1. As a way of developing some images of classroom discourse, read *What's All This Talk About "Discourse"?* by Deborah Ball. (Note that this article will be discussed again in Session 6.)

2. Observe two teachers in your school or district teaching a mathematics lesson. Choose teachers who are not being evaluated for professional status and who are not encountering any problematic teaching issues. Use the Pre-observation Conference Questions to help you prepare for the observations. Use the Intellectual Community Observation Guide in your observations.

3. Write up a short summary of each observation, addressing as many of the questions in the Intellectual Community Observation Guide as are relevant. Start with your overall impressions of this classroom. Then describe the intellectual community. Conclude with any questions or issues you are puzzled about.

Pre-observation Conference Questions

In order to help you make sense of what you will be seeing when you do your classroom observations, plan to meet with the teacher prior to the observation and ask the following questions:

1. What topic will you and your students be working on in this lesson?

2. What do you plan to do in this lesson? (e.g., the origin and structure of the lesson, and so on)

3. What do you hope to accomplish in this lesson?

4. What mathematical ideas are embedded in this lesson?

5. What have you and your students been working on prior to this lesson?

6. How does this lesson fit into your overall goals for the year?

7. Are there students who have special issues in the class?

Lenses on Learning, SESSION 3

Intellectual Community Observation Guide

Students		
Focus Question	Conjectures	Evidence from Classroom
How are students showing respect for one another's ideas?		
• What are students doing to show that they are listening to other students?		
• How willing are students to share their ideas even if they know that they aren't correct?		
• How attentive are students to one another's ideas?		
How do students use each other as resources as they make sense of mathematical ideas?		
• Are they building on each other's mathematical ideas?		
• Are they asking each other questions related to mathematical ideas?		
What evidence beyond raised hands do you have that students are engaged?		

© Education Development Center, Inc.

36 ♦ Session 3

Intellectual Community Observation Guide

Teachers		
Evidence from Classroom	Conjectures	Focus Question
		How does the teacher support students in showing respect for one another's ideas?
		• How does the teacher set and maintain norms of interaction for discourse?
		• How does the teacher support such practices as: • attentive listening • question-posing about mathematical ideas • provisional thinking
		How does the teacher set the tone for students to see each other as resources for mathematical thinking?
		• Does the teacher invite students' tentative thinking?
		• Does the teacher invite students to build on one another's ideas?
		What interventions does the teacher make to ensure that students' engagement has a focus on mathematical ideas?

READING 3

What's All This Talk about "Discourse"?*

Deborah Loewenberg Ball
Susan N. Friel

Despite its title, the *Professional Standards for Teaching Mathematics* (NCTM 1991) should not be read as a set of prescriptions about how to teach. The document will not deliver on such expectations, not because it fails but because no document can prescribe good teaching. No set of standards can be expected to stipulate what teachers should do. The potential of the *Professional Teaching Standards* rests instead in its use as a set of tools with which to construct productive conversations about teaching. It should be viewed as a resource with which to build teaching rather than as a measuring stick by which to judge teaching. With new ideas about things to pay attention to in our classrooms, to ask ourselves, to wonder about, we would have increased power to analyze and improve our teaching—alone and as members of a wider community of educators. In this article I explore possible outcomes of using the *Professional Teaching Standards* in such ways.

Discourse in the Classroom

The *Professional Teaching Standards* calls unprecedented attention to the "discourse" of mathematics classrooms, as embodied in three standards: Teacher's Role in Discourse, Students' Role in Discourse, and Tools for Enhancing Discourse. An unfamiliar term to many, *discourse* is used to highlight the ways in which knowledge is constructed and exchanged in classrooms. Who talks? About what? In what ways? What do people write down and why? What questions are important? Whose ideas and ways of knowing are accepted and whose are not? What makes an answer right or an idea true? What kinds of evidence are encouraged or accepted?

The discourse of a classroom is formed by students and the teacher and the tools with which they work. Still, teachers play a crucial role in shaping the discourse of their classrooms through the signals they send about the knowledge and ways of thinking

---

Deborah Ball is associate professor of teacher education at Michigan State University. On a daily basis, she also teaches mathematics to Sylvia Rundquist's third-fourth-grade combination class at Spartan Village Elementary School. East Lansing, MI 48823. She conducts research on teaching and learning to teach mathematics and is especially interested in problems of changing practice. Ball chaired the Working Group on Mathematics Teaching for the new Professional Standards for Teaching Mathematics.

The Editorial Panel welcomes readers' responses to this article or to any aspect of the *Professional Standards for Teaching Mathematics* for consideration for publication as an article or as a letter in "Readers' Dialogue."

---

*From *Arithmetic Teacher*, November 1991, pp. 44–48. "What's All This Talk About 'Discourse'?" by Deborah Loewenberg Ball and Susan N. Friel. Copyright ©1991 Association for Supervision and Curriculum Development. Reprinted by permission. All rights reserved.

and knowing that are valued. For example, suppose a student claims that in tossing a penny, heads is a more likely result than tails. He explains that in ten throws, he got seven heads and three tails. He says that this outcome shows that it is "easier to get" heads. How might the teacher respond to this statement? Knowing herself that heads and tails are equally likely, she might tell him that this conclusion isn't right and explain that the outcome he noted was simply what happened that time. Or she might ask other students to respond to his assertion, for across the entire class the cumulative results of tossing a penny will probably turn up approximately half heads and half tails. Or she might observe that ten throws is not very many and suggest that he gather more data himself to see what happens. In each of these alternatives, different messages are sent about the usefulness and validity of the student's experience with coin tossing. Each conveys different implications about the role of experimentation in constructing mathematical knowledge, and each may have different effects on the student's view of mathematical justification. This small example illustrates how influential teachers' interactions with students are in shaping norms of knowing and thinking. These interactions, in turn, influence students' ways of knowing. The norms that students and teachers come to share deeply affect the potential of the classroom as a place for learning.

Without explicit attention to the patterns of discourse in the classroom, the long-established norms of school are likely to dominate—competitiveness, an emphasis on right answers, the assumption that teachers have the answers, rejection of nonstandard ways of working or thinking, patterns reflective of gender and class biases. For example, in many mathematics classrooms, answers have traditionally been right because the teacher says so or because the teacher and the student together decipher what "they" (the textbook authors) "want." Even with careful attention to patterns of classroom discourse, traditional norms will underlie the interactions of students and teachers. Consider the way in which right answers are treated in a mathematics class. Suppose students are solving the problem "What is two-thirds of nine?" and a student gives the answer, "Six." The teacherly reflex is to hear it as a "right answer" and to (a) move on; (b) praise the student; or (c) agree and repeat the answer for the benefit of the rest of the class. Even disposed to ask students to explain their answers, the teacher may ask. "How did you *know* that?" In listening to myself in my own classroom, I realized that using the word *know* seems simply that the student's answer is right. I heard myself saying, "How do you *know*?" when I agreed with what a student said and asking, "Why do you think so?" or "How did you get that?" when I did not. Subtly, I was probably giving my students clues about the "correctness" of their ideas—clues they were likely picking up.

When we hear right answers simply as representing understanding, we miss opportunities to gain insight into students' thinking. A student could get six as the answer to "What is two-thirds of nine?" by using the following reasoning: two-thirds mean two groups of three, which is six. Although interesting, this is not the conventional interpretation of two-thirds. Similarly, when we hear wrong answers as representing a lack of understanding, we also miss the opportunity for valuable insights. Some wrong answers are produced by simple errors, whereas others represent well-developed concepts or ways of thinking—some productive and some less so. For example, a student in my class is convinced that two-thirds was less than one-sixth. I could have interpreted this perception as a problem with her ability to compare fractions with unlike denominators or suspected that she was thinking that $\frac{1}{6}$ was more than $\frac{2}{3}$ because it has a "6" in the denominator, whereas $\frac{2}{3}$ has only a "3" (and 6 is more than 3). Instead, she was reading $\frac{2}{3}$ as "you have three parts and you take two of them away" and $\frac{1}{6}$ as "you have six parts and you take (only) one of them away." Her reasoning about fractions was complicated with subtraction concepts that led her to believe that more was "left" in one-sixth than in two-thirds. Had I not explored this student's answer with her, neither I nor the other students would have understood the thinking that led to her conclusion.

The classroom environment, or culture, that the students and teacher construct affects the discourse in some important ways. The environment shapes how safe students feel, whether and how they respect one another and themselves, and the extent to which serious engagement in mathematical thinking is the norm. Are students' voices and thinking valued by the teacher and by other students? What norms are established for the exchange of ideas? How are disagreements expressed and handled? How much risk is involved in being wrong? To what extent can every student participate and learn in the class?

Our Professional Discourse as Teachers

Working to become as skillful as possible in our classrooms requires us to learn to see more and more broadly and deeply. We must examine the language of our work with students, reflect on the direction and tone of class discussions, consider the time we allow students to explore and investigate—all these endeavors are critical in achieving the discourse we foster. Facilitating worthwhile learning seems very much a matter of orchestration—of eliciting and interweaving multiple voices, threads, themes, and tones. Examining alternate perspectives on the intellectual and social classroom environment can enhance the virtuosity of our work. This is no easy task.

Perhaps the most valuable part of my experience, working on the *Professional Teaching Standards* was what it contributed to growth in my own teaching. As we debated and wrote, listened to others and revised, I gained new ways of looking at what I was doing as a teacher. Many specific people stand out in my memory. When I think about issues of classroom discourse, I think particularly of Susan Friel. Friel asked me tough questions about the kinds of words I used to help my students learn to argue with one another about mathematics. She made me think hard about how I cast the tone of the discourse in my classroom. Specifically, she made me wonder about how students felt when other students said things like, "I challenge you." Was challenge an appropriate word to be using? Did the students sound aggressive? How safe was the environment? Was the attitude I was encouraging somehow less inviting to girls than to boys? Friel's questions made me watch and listen more closely, and I changed some of the ways in which I taught my students to talk with one another about mathematical ideas.

Let me illustrate. In Figure 1 I describe a discussion that took place in my third-grade classroom last year. My purpose is to use the discussion as a context for considering issues central to the shaping of the discourse within the classroom and to offer an example of the kind of discourse about teaching that I think the *Professional Teaching Standards* can promote.

I am not telling this story to set myself up as an exemplar of "good teaching."

Quite the contrary. The snippet is in many ways quite ordinary, and it brings up tough questions that we all face many times a day. As you read, consider what is going on, consider my comments and questions to myself, and consider the issues you would take into account in orchestrating this lesson yourself.

My commentary on the lesson segment is of course incomplete. I did not raise all possible considerations nor reflect on all the things that were on my mind. I did not include others in this particular conversation with myself. Readers who examine with me this excerpt from my teaching will notice different things than I did. They will have different concerns and questions. No single issue is the essential one; none is definitively inappropriate. The traditional isolationist culture in teaching—that everyone has to find his or her own style, that admitting to reaching an impasse or having a hard time is tantamount to an admission of incompetence has been a crippling aspect of our work as a community of educators. At the same time, although we know better, we seem to talk as though "a right way" exists to motivate students, to teach place value, or to respond to certain kinds of questions from students. On one hand, then, we have pretended that we have nothing to learn from one another. And on the other, we have pretended that teaching is simple and straightforward.

Establishing and maintaining patterns of discourse, fostering the environment of the classroom—these are not matters of right or wrong. But by articulating our thinking and concerns—to ourselves and to others—we can increase our own professional skills. By raising new questions and issues that shape the ways in which we see and think about our classrooms, we can enhance our orchestration of interaction in our classes in ways that can contribute to the kinds of learning outlined in the *Curriculum and Evaluation Standards for School Mathematics* (NCTM 1989). This is the role that I hope the *Professional Teaching Standards* can play—to give us tools for examining and building our teaching as individuals and as a community.

Conversations of this sort—with ourselves and with a wider range of others—can help us as teachers develop our sensitivity to, and repertoire for, structuring the discourse of our classrooms. Although it cannot prescribe what any one of us does from moment to moment in our classrooms, the *Professional Teaching Standards* can provoke us to work alone and with one another in ways that we have not typically done. We could learn to talk reflectively and analytically about what we do and what we think and about the struggles and dilemmas with which we contend on a daily basis. To make this change will require some substantial changes in our professional discourse—in what we talk about with one another and in what ways. Such change requires us to rethink our assumptions about what counts as evidence for believing or doing something in teaching and to let one another behind the proverbial classroom door—to explore one another's practices, to raise hard questions, to help one another grow.

References

National Council of Teachers of Mathematics, Commission on Professional Standards for Teaching Mathematics. *Professional Standards for Teaching Mathematics.* Reston, Va.: The Council, 1991.

National Council of Teachers of Mathematics, Commission on Standards for School Mathematics. *Curriculum and Evaluation Standards for School Mathematics.* Reston, Va.:

Figure 1.
A third-grade-class discussion and teacher's mental discourse

It is a warm mid-May afternoon. My third graders have been working on the problem, "Three-fourths of the crayons in Mrs. Rundquist's box of a dozen crayons are broken. How many unbroken crayons are there?"

A boy named Sean offers to show his solution. "It would be four," he assert as he comes up to the chalkboard. He draws twelve sticks to represent the twelve crayons and marks off groups of four crayons:

He explains, "Well, I, um, counted these and I got, I went, 'One, two, three, four,' and I put a line down. So it's . . . then I went, 'One, two, three, four'; and I put another line down and I add them up and it's eight; and I put another line, 'One, two, three, four.' And that was twelve." He finishes. "Why . . ." I begin to ask, but Sean interrupts, changing his mind. "A quarter wouldn't be *that*." He erases the lines, "Because, um, because that's a *third*. There's only three groups. There's supposed to be *four* groups." Sean draws lines to mark off four groups of three crayons:

He explains, "Because it's three-fourths, that's what I said, it's three-fourths; so three crayons is a fourth, so three, and that's a fourth, that's a fourth, and that's a fourth, so that's three-fourths."

Riba, waving her hand, disagrees. She says that one-fourth should have four crayons in the group—like Sean had presented it at first. "This is what I think: three-fourths is like, um, three groups of four."

I ask for other students' reactions. Ofala says she agrees with Sean. "I think he's right because he's taking the three, like separating the three groups plus the one group he didn't circle."

I probe: "Why does he have three in every group instead of four in every group?"

Sean says that if it was three groups of four, "this should have four in each one, and it would be

I have a very diverse class, with several students who are just learning English. Is this problem one that all my students can engage in reasonably?

I always wonder about whom to call on to start off our discussion of a problem. Who should have the floor first and why?

Accepting drawings like this one seems important to expand the tools that students can use to think with as well as to express their thinking.

I am glad to see that Sean expects that part of showing his solution is to explain what he did and what he was thinking.

I am always interested when students figure out for themselves that something doesn't make sense. How to set up the environment so that students feel comfortable about changing their mind is a big concern, since in school being "wrong" is traditionally something to hide or to be ashamed of.

Should I say something here? Many teachers would praise him for his explanation and for figuring out and revising his answer. I would like, though, for the students to come to rely less on me for confirmation and more on themselves.

Should I ask Riba to say first what she thinks of Sean's explanation, partly to help her learn what it means to show respect for his ideas and partly to make sure she has really understood what he

sixteen." (In other words, if three groups of four was the answer to three-fourths, it would have to be three-fourths of sixteen, not of twelve.)

He erases the extra four lines and turns to Riba. Pointing at the drawing, he says, "These aren't fourths, these are thirds because there's three groups and that makes them a third." Keith raises his hand and explains that what Riba is saying is that one-fourth means "one group of four." Riba nods. Sean turns to the class, and the following dialogue ensues:

Sean: Let's take a vote! How many people, um, think that my answer is correct raise their hand, and how many people think Riba's answer is correct raise your hand.

Teacher: Why would that be a good idea? What would that do if we saw that? Why would we want to know that?

Sean: That would prove it.

Students: No! Not!

Teacher: Keith, you're shaking your head. Why wouldn't that prove it?

Keith: Just because, like, just because somebody agrees with another person doesn't mean that they're right.

Betsy: I have an example of why voting doesn't work. When we were talking about zero, if it was an odd or even, a whole lot of people said that it was an odd; but then afterward we figured out that it was even, and voting didn't help us know if it was odd or even because the answer was opposite than what people had voted.

Teacher: So how did we change our minds, then, if the voting doesn't work?

Betsy: Because the people found out patterns and the number line and they figured out that no, zero must not be an odd because when it goes up there it goes odd, even, odd, even, odd, even; and so when you had an odd number like one and then you have zero, zero must be even because that's the way it is.

was saying?

The teacher's role in orchestrating discussion is so

hard! Should I clarify what the "4" means in "$\frac{3}{4}$" or

should I let the other students say what they think?

Here I decided that Riba's idea that $\frac{3}{4}$ implies groups

of four was worth probing further. Does it make sense to do this here—to open up the discussion to a "wrong" idea when it seems to be moving in a "right" direction?

Teacher: Anybody else want to comment on this before we go back to our problem of fourths and thirds? Mei?

Mei: I don't think it would work, but it would be fun to see how many people agree with him because maybe some people would come up with some other idea.

Teacher: So, you'd be curious just to know what people are thinking?

Mei: Yeah

Sean: I agree. But that's a really hard question that Riba is asking, but why shouldn't there be four groups of, um, three.

I suggest that we return to trying to interpret what three-fourths might mean. Betsy volunteers that she has an idea.

The situation seems to be becoming competitive—a situation in which Riba's ideas are pitted against Sean's and in which one of them will be a "winner." Voting has been a big part of their experience in settling group matters. Yet figuring out what makes sense in this situation does not seem to be a matter of democratic vote. Should I just explain this idea to him? I decide to ask him what he is thinking.

What is Sean's notion of what makes something right or true in mathematics?

This seems to be a good example from their shared experience that may help the students understand why knowledge isn't simply legislated.

With a sudden start I recall how often I have "polled" the class—not to settle matters of disagreement but to give myself and the students some information about the distribution of ideas in the group. I do not think of this approach as "voting"—that is, as a means to determine the correct answer. But this distinction is subtle, and I do not know what the students may be thinking about the role of voting.

Betsy: I'm thinking about what's a fraction that you know is true. A fraction that you know, that we already agreed . . . just wait . . . (pauses, thinking of an example) Okay, yesterday people agreed on half of twenty-four was twelve, right? (to Riba) Do you agree with that? Half of twenty-four is twelve? Well, if we put twenty-four lines, we don't circle two in each group, do we? We went, "One, two, three, four, five, six, seven, eight, nine, ten, eleven, twelve." Then we cut it right there and we circled this half, and that would be half.

See, we have, we have two groups here, right? This (pointing to the "2" in "$\frac{1}{2}$.") means the groups.

Why is Betsy doing so much of the talking today? Should I do something here to elicit the ideas of other students, or should I call on her and see what she says and then go from there? It seems that on different days, different students are more active than

SESSION 4

Observing for Content: Listening to Children's Ideas About Fractions

In this session you will continue to develop a new "eye" for mathematics classrooms with a focus now on the mathematics content of the lesson. You may be accustomed to paying only cursory attention to the mathematics content in the lessons you observe, taking note of the topic of instruction and the instructional activity the teacher proposes. In this session, you will consider the importance of looking beyond the broad topic to the mathematical ideas that fall within that topic in order to grasp more fully the mathematical content of the class.

In standards-based classrooms both the mathematics under investigation and the teacher's knowledge of it are centrally important. As an observer and teacher supervisor, you need to attend to the mathematical thinking that students are engaged in and to the ways that the teacher helps students develop their thinking.

These ideas will be the focus of investigation and discussion in this session and the next. The Math Content Observation Guide will provide a frame for you to use as you look at the mathematical concepts being explored and the teacher's knowledge of the math content in the videotaped clips or in actual classroom observations.

Approaches to Solving Math Problems and Magical Hopes

1. The first homework assignment for Session 4 is to read an excerpt from *Young Mathematicians At Work: Constructing Fractions, Decimals, and Percents* by Catherine Twomey Fosnot and Maarten Dolk (Reading 4).

   This excerpt describes the approaches two pairs of students in a fourth/fifth grade classroom used to solve a fair-sharing problem.

   As you read the excerpt, work through the ways the students approached the problem. Also note what mathematical ideas seemed to be difficult or puzzling for students.

2. In the next session we will consider the Math Content Observation Guide. In order to be ready to view videotaped clips from this perspective, it is important to distinguish between the mathematical *ideas* that students in a class are working with and the ways in which those ideas might be represented in instruction—with manipulatives, diagrams, or numbers.

   As a way of thinking about the relationships between mathematical ideas and their representations, please read *Magical Hopes: Manipulatives and the Reform of Math Education* by Deborah Ball (Reading 6). Identify three ideas in this article that are especially interesting to you and write at least a paragraph about each.

READING 4

Approaches to Solving a Fair-sharing Problem*

Join us in Carol Mosesson's fourth/fifth-grade classroom, in New Rochelle, New York. She is telling her students about a dilemma that occurred in her class the previous year and how she wants to be sure it doesn't happen again.[1]

"Last year," Carol explains, "I took my students on field trips related to the projects we were working on. At one point, we went to several places in New York City to gather research. I got some parents to help me, and we scheduled four field trips in one day. Four students went to the Museum of Natural History, five went to the Museum of Modern Art, eight went with me to Ellis Island and the Statue of Liberty, and the five remaining students went to the Planetarium. The problem we ran into was that the school cafeteria staff had made seventeen submarine sandwiches for the kids for lunch. They gave three sandwiches to the four kids going to the Museum of Natural History. The five kids in the second group got four subs. The eight kids going to Ellis Island got seven subs, and that left three for the five kids going to the Planetarium." As she talks, she draws a picture (see Figure 1.1) on chart paper of the context she is developing. "Now we didn't eat together, obviously, because we were all in different parts of the city. The next day after talking about our trips, several of the kids complained that it hadn't been fair, that some kids got more to eat. What do you think about this? Were they right? Because if they were, I would really like to work out a fair system—one where I would know how many subs to give each group when we go on field trips this year."

Carol is introducing fair sharing—a rich, real context in which her students can generate and model for themselves mathematical ideas related to fractions. When children are given trivial word problems, they often just ask themselves what operation is called for; the context becomes irrelevant as they manipulate numbers, applying what they know. Truly problematic contexts engage children in a way that keeps them grounded. They attempt to model the situation mathematically, as a way to make sense of it. They notice patterns, raise conjectures, and then defend them to one another.

---

*From *Young Mathematicians at Work: Constructing Fractions, Decimals, and Percents* by Catherine Twomey Fosnot and Maarten Dolk. Copyright ©2002 by Catherine Fosnot and Martin Dolk. Portsmouth, NH: Heinemann, a division of Reed Elsevier, Inc. Reprinted by permission of the publisher. All rights reserved.

[1] Although the investigation was embellished and situated in the context of field trips by Carol Mosesson, the kernel of the activity—fair sharing of submarine sandwiches—comes from Mathematics in Context, Encyclopedia Britannica Educational Corporation.

"Turn to the person you are sitting next to and talk for a few minutes about whether you think this situation was fair," Carol continues.

"It's fair," several children comment. "It's one less sandwich than kids each time. It's three subs for four kids, so when there are five kids, they gave them four. For eight kids, they gave them seven."

"Yeah, but it wasn't fair for the Planetarium group. They had five kids and they only got three subs." Jackie is adamant as she comes to their defense. "The Museum of Modern Art group had five kids, too, and they got four sandwiches!"

"But you could just cut them in different pieces, like fourths or fifths," John offers.

"But the pieces would be different sizes. It's not fair."

"What do you think about her argument, John? Did everyone get the same amount?" Carol inquires, interested in whether John thinks the pieces would be equivalent. He shakes his head, acknowledging that the pieces would not be the same size. Another student raises his hand "Michael?"

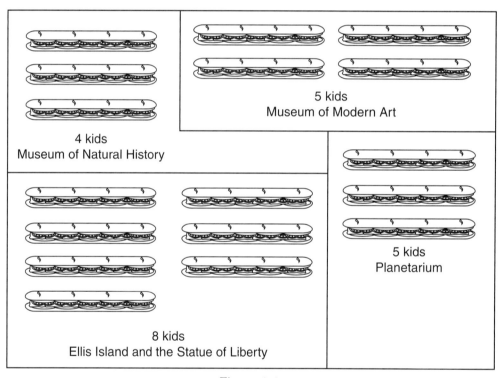

Figure 1.1

"But the others *are* all the same." As Michael advances this common misconception, several of his classmates nod in agreement." You keep adding one child and one sub. Three for four is the same as four for five."

"If the Planetarium group had another sub it would all be the same," Aysha says, verbalizing what most are thinking.

Carol has succeeded in making their initial ideas visible. Now she wants to create disequilibrium. "Well, let's investigate this some more. I suggest we work with our math workshop partners and check out two things." She writes the question down on a large piece of chart paper as she talks. "How much did each child in each group get, assuming the subs were all shared equally in each group? And which group got the most?"

Several students continue to voice their agreement with Aysha and Michael. "I think they all got the same except for the Planetarium group."

"Well, let's check it out and see. Let's investigate a bit, prove your ideas, and we'll have a math congress after we've worked on it some. Use any materials you want, or make drawings to help you prove your thinking. We'll share our thinking when everyone feels he or she has had enough time to work on this."

The children set to work investigating the problem. Some take out Unifix cubes; most draw the subs and show how they would cut them. Carol moves from group to group, ensuring that everyone is busy and clear about the problem. She sits down with Jackie and Ernie for a moment.

"There were three subs for four kids," Jackie explains, "so we cut two subs in half, then the last in fourths. So everybody in that group got one half plus one fourth."

Ernie, her partner, explains their work for the Ellis Island group. "And we're doing the same thing here. See—we're giving each person a half sub first. That is four subs. One sub we're cutting into eight pieces, and the other two into fourths."

"So how much does each person in the group get?" Carol inquires.

"One half plus one eighth plus one fourth."

"And you're going to continue like this with the Planetarium and the Museum of Modern Art groups?" Jackie and Ernie nod yes, Carol notes their strategy: they are making unit fractions—fractions with numerators of one.

Nicole and Michelle, whom she visits next, have worked on the three-subs-four-kids situation very similarly. But for the four-subs-five-kids situation they have changed their strategy. Michelle explains, "We divided these subs up into fifths, because there were five kids."

Carol sees that each of the four subs is drawn and cut up into fifths. "So how much of a sub did each child get?" she asks.

"One of these, one of these, one of these, and one of these." Nicole points to a fifth from each sub "That's four fifths of a sub for each kid because it is four times one fifth."

"Oh, that's interesting, isn't it? Four subs for five kids ended up being four fifths of a sub." Carol attempts to point up the relationship but Nicole and Michelle seem uninterested, and they go back to using unit fractions for the seven-subs-eight-kids problem. Carol debates whether or not to suggest that they try the same strategy they used for the four-subs-five-kids situation but decides against it. They are very involved in the context and they (as well as Jackie and Ernie) are cutting up the subs in ways that make sense *within the context*. They can imagine subs cut up into halves, fourths, or even fifths. These are reasonable sizes. Eighths are not, unless they become necessary at the end to share a last sub fairly. The realistic nature of the context enables children to realize what they are doing, to check whether it makes sense.

Approaches to Solving a Fair-sharing Problem ♦ 49

Working within a context also develops children's ability to make mathematical meaning of their own many lived worlds. There is much to be investigated yet. Asking children to adopt multiplication and division shortcuts too soon may actually impede genuine learning. As the well-known mathematician George Polyá (1887–1985) once pointed out, "When introduced at the wrong time or place, good logic may be the worst enemy of good teaching." Historically (see the in-depth discussion in Chapter 3), unit fractions were used for a very long time before common fractions were accepted. As the children explore fair-sharing situations with unit fractions, they will have experiences comparing and making equivalent fractions. These experiences will bring up some big ideas and support the development of some powerful mathematical strategies. Let's return to Jackie and Ernie at work to witness this process.

Jackie has drawn four subs, and she and Ernie are deciding how to share this amount fairly among five kids—the ones who went to the Museum of Modern Art. "Let's give everyone a half again first. That leaves one and a half subs left."

"So let's cut the whole one up into fifths." Ernie draws four lines, making five equal parts, as he talks.

"Okay, so now every kid has one half plus one fifth," Jackie continues, "and now we have to cut up the last half into fifths. So it's one half plus one fifth plus—*one fifth?* That can't be—these are just slivers." She points to the small pieces resulting from cutting up the half sub. "These slivers are only about half the size of those fifths. So what do we call these?"

Jackie and Ernie are struggling with a big idea. To resolve their puzzlement they must grapple with the heart of fractional relationships. One fifth of a half sub is smaller than one fifth of a whole sub. What is the whole? And what is the equivalence of this piece to the whole?

"You're right—two of these fifths [the slivers] are about the same as one of those fifths," Ernie ponders. "So maybe it's a half of a fifth. But what is that?"

"We know it is a fifth of a half of a sub—"

"Yeah, but it is also one half of a fifth—see." Ernie points out the relationship of the sliver to the fifths that resulted when one sub was cut into five equal pieces. He is noticing that $\frac{1}{2} \times \frac{1}{5} = \frac{1}{5} \times \frac{1}{2}$. This relationship is a specific example of the commutative property of multiplication, although for him at this point it is probably not generalizable.

"Well, if it's a half of a fifth, then it must be a tenth," Jackie offers.

Ernie ponders her suggestion and then finds a way to prove it. "Oh, I get it—look. If this half is cut into fifths to make the slivers, then the other half could be too. And then—you're right—there's ten pieces in the whole sub. So one fifth of a half is one tenth."

"So now the Planetarium group is easy. Everybody gets a half plus one tenth!" Jackie draws the subs for the last scenario, showing the fair sharing (see Figure 1.2). "Now all we have to do is compare them!"

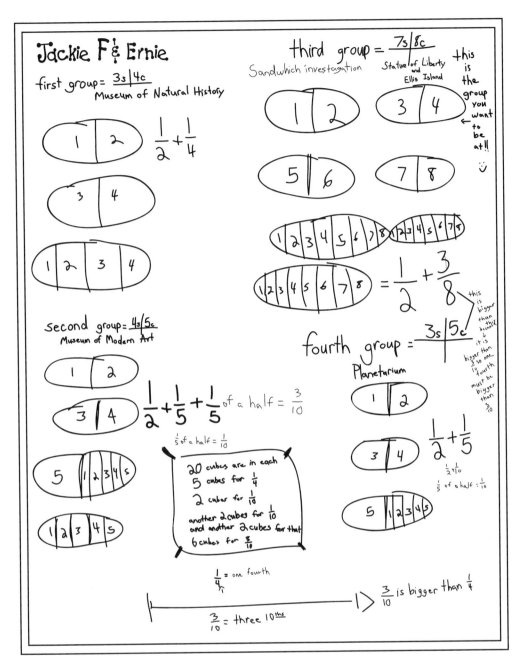

Figure 1.2

David Hilbert (1897), in his book, *Report on Number Theory*, wrote, "I have tried to avoid long numerical computations, thereby following Riemann's postulate that proofs should be given through ideas and not voluminous computations." As they set out to compare the four scenarios and prove which group got the most, Jackie and Ernie could add up the fractional parts and make common denominators, but they don't. Working with unit fractions has allowed them to hip upon a brilliant strategy to make the problem simpler.

"Everybody, in every group, got a half. So we can eliminate that," Jackie suggests.

"So all we have to do is compare the remainders," Ernie finishes her thought. "So we have one fourth for the Museum of Natural History group, one fifth plus one tenth for the Museum of Modern Art group, one fourth plus one eighth for the Ellis Island group, and one tenth for the Planetarium group.

"The Ellis Island group got one eighth more than the Museum or Natural History group."

"And the Museum of Modern Art group got one fifth more than the Planetarium group."

"But how do we compare the others?"

Ernie grabs some Unifix cubes, "I have an idea! What's a good number to use?" He ponders for a moment then clarifies for Jackie, "I'm trying to think of a number that we could use to build the subs—a number that five and four and ten would go into—"

"How about twenty?" Jackie offers.

"Yeah—that's good." Ernie counts out twenty cubes and builds a stack with them to represent a sub. "Okay—so the fourth is five cubes, the fifth is the same as two tenths. . . . and one tenth more, that's three tenths." As they begin to combine amounts and use twenty as a common denominator, they make a first attempt at generating common fractions. Jackie continues, "Ten times two is twenty, so two, four, six—three tenths is the same as six twentieths. That's more than five twentieths!"

At first the eighths stump them: eight does not go into twenty evenly. But Ernie returns to their drawing. Using his fingers, he points to each eighth and says, "Two of these make a fourth—so the Ellis Island group got three eighths"

Jackie excitedly finishes the comparison. "Eighths are bigger than tenths. We're comparing three eighths to three tenths. If you cut up a sub into tenths, those pieces are smaller than if you cut it up into eighths. The threes are the same on the top," referring to the numerators, "so the Ellis Island group got the most—three eighths of a sub!"

"Plus the half," Ernie reminds her. "And the Planetarium group got the least. One tenth is the littlest because if you divide a sub into fourths or fifths or tenths—the more pieces you have the smaller the piece is! They only got a half plus one tenth!"

READING 5

Fraction Tracks Transcript[1]

William H. Lincoln School in Brookline MA

Teacher:	This morning we quickly looked at a game we are going to play and I wanted to quickly review the game with you and then I am going to allow you to start playing the game.
Teacher Voiceover:	My name is Hilory Paster. I teach grade 5 at the Lincoln School in Brookline Massachusetts.
Teacher:	We talked about that this is the Fraction Track activity and for this activity you are going to be moving your chips over from the zeros side to the ones side.
Teacher Voiceover:	Today's lesson students had to line up markers along fraction lines and they were asked to pull a card from the fraction pile and find ways to move chips from one end of the board to the other end of the board to reach one whole.
Teacher:	And the card we said we were going to talk about this morning was $\frac{8}{10}$. Did anybody think of a way that you could move you piece across to the other side if our card was $\frac{8}{10}$?
Rebecca:	Instead of using $\frac{8}{10}$ you could use $\frac{2}{5}$, no $\frac{4}{5}$ and you could go to $\frac{4}{5}$ or $\frac{8}{10}$.
Teacher:	So I could go to $\frac{4}{5}$ or $\frac{8}{10}$ so one or the other? How did you know that $\frac{4}{5}$ was a good move?
Rebecca:	Because $\frac{8}{10}$ and $\frac{4}{5}$ are equivalent.
Teacher:	Ok.
Rebecca:	So both of them would work.
Teacher:	Is there a way that I could move 2 pieces rather than just one because in the way that you described it, Rebecca, I have to use either on the fifths line or the tenths line but I couldn't use both. Is there a way I could start again and use both.
Student:	You can first go to $\frac{4}{10}$.

---

[1] From *Teaching Math: A Video Library, K–4* and *Teaching Math: A Video Library 5–8*. Funded by Annenberg CPB and produced by WGBH Boston. Use by permission of the producer.

Teacher:	$\frac{4}{10}$, ok.
Student:	And that would be halfway. And then you can go to $\frac{4}{5}$.
Teacher:	$\frac{4}{5}$? Tell me what you are thinking about that.
Student:	Since they're both equivalent and you go halfway
Teacher:	But now I've moved $\frac{4}{10}$. Can I move the whole $\frac{4}{5}$ now?
Student:	$\frac{2}{5}$?
Teacher:	$\frac{2}{5}$? Because then if I move $\frac{2}{5}$ and $\frac{2}{5}$ plus $\frac{4}{10}$ would then equal how much?
Student:	$\frac{8}{10}$.
Teacher Voiceover:	The students know a great deal about equivalencies. We have done a number of activities which led them to the point where they could use them regularly and more frequently. We started by having them draw fraction cards in which they would have a fraction such as $\frac{2}{4}$ and they would draw a picture to model that and then they would try to find other fractions that equaled the same amount. Then they strung them on a clothesline, ordering the fractions. We started keeping a list in the classroom of equivalencies that they knew and were coming across of so we would always be referring back to this list. So after a number of days working with them, they started to become second nature to them, and they were able to move around between them.
Teacher:	What I want everybody to do now, is you are going to start the game. You only need between you and your partner, one fraction track. I also want to remind you that you have to keep track of your moves. So on your fraction track strategy sheet, I want you to write down the fraction card that you pulled and 2 to 3 possible moves that you could use before you decide to move, 'cause the goal of the activity is for you as a team to move your pieces across to the one whole and in order to do that there is more than one way you can get there, so in order to do that I want you to think of the best move.
Teacher:	So one move you are saying Sean is what?
Sean:	I could just do $\frac{4}{10}$ or I could do $\frac{2}{10}$ and $\frac{1}{5}$.
Teacher:	How does he know that $\frac{2}{10}$ and $\frac{1}{5}$ would be something that would work?
Student:	Because $\frac{1}{5}$ is half of $\frac{4}{10}$ and $\frac{1}{5}$ is equivalent to $\frac{2}{10}$.
Teacher:	Excellent. Write down those two strategies that Sean has and I am going to deal with the little problem over there and then I am going to come back and see what you've come up with. If there's another . . . there might not be another one but there might be another one that you could do.
Sean:	You could do $\frac{2}{5}$ and . . . .

Student:	Yeah because $\frac{2}{5}$ and . . .
Sean:	I think we should just double that.
Student:	Yeah, 'cause its equivalent.
Student:	So I gotta write down the two things so . . . $\frac{2}{10}$ and $\frac{1}{5}$ and $\frac{4}{10}$ and $\frac{2}{5}$.
Student 1:	What card are we doing?
Student 2:	$\frac{6}{10}$
Student 1:	$\frac{6}{10}$, ok uh $\frac{6}{10}$ plus $\frac{2}{5}$?
Student 3:	$\frac{6}{10}$ plus $\frac{2}{5}$? Well you are trying to get $\frac{6}{10}$ as a total so if you do $\frac{6}{10}$ plus $\frac{2}{5}$ that will be more Than $\frac{6}{10}$, wouldn't it?
Student 1:	$\frac{1}{10}$ plus $\frac{5}{10}$.
Student 3:	And $\frac{1}{10}$ plus $\frac{5}{10}$ equals what?
Student 1:	$\frac{6}{10}$.
Student 3:	Ok.
Tamara:	The last time I got $\frac{2}{6}$ and now I just picked up $\frac{8}{8}$.
Teacher:	alright.
Student 2:	She knows she can go to $\frac{8}{8}$ but she has to start off from there and we can't tie in sixths with eighths so we don't know what to do.
Teacher:	So is your question that you have $\frac{8}{8}$, but on your eighth line, you're on $\frac{2}{8}$.
Tamara:	No . . . . fades.
Teacher Voiceover:	One of the areas that they were having difficulties with was finding agreement with what move they wanted to make. I think that one student might have seen an easier move but another student saw one that was a little more complicated and going around the room and giving them ways in which to show their thinking to each other helped them to be able to negotiate about which move they wanted to make.
Teacher:	I can move on my four, so what kind of move would I make there?
Tamara:	You can go the rest of this with $\frac{6}{8}$.
Student 2:	And then take $\frac{2}{8}$ which would be $\frac{1}{4}$.
Teacher:	I could do that.
Tamara:	So that would go up to $\frac{2}{4}$.
Teacher:	Why $\frac{2}{4}$?
Tamara:	Because $\frac{6}{8}$.
Teacher:	OK you have $\frac{6}{8}$ and what did you say you could move next?
Tamara:	You'd need to move 2 more eighths

Teacher:	I could... ok, so that's plus two eighths and that would equal $\frac{8}{8}$. You were giving me the suggestion of moving on my fourths line. Can I do that? $\frac{6}{8}$ plus what equals $\frac{8}{8}$?
Student 2:	$\frac{1}{4}$.
Teacher:	Tell Tamara how you got that $\frac{1}{4}$ because I'm not sure she knows.
Student 2:	Because $\frac{6}{8}$.... is $\frac{6}{8}$, but $\frac{1}{4}$ is $\frac{2}{8}$.
Teacher:	And $\frac{1}{4}$ is the same as $\frac{2}{8}$, so he's doing this. So now you can take this chip off and put a new one on here and you can leave that one.
Student 1:	I can't move that.
Student 2:	Well is there anything with that..?
Student 1:	Wait, yeah, yeah, oh I can't do that.
Student 3:	$\frac{2}{10}$ is equal to $\frac{4}{5}$.
Student 4:	Is that past four?
Student 3:	I mean
Student 2:	I can't move at all.
Student 1:	So do you just skip your turn if you can't move? (picks another card) $\frac{2}{10}$.
Student 2:	$\frac{2}{10}$. That's $\frac{1}{5}$. Now you have to skip your turn too, I guess. (picks a card) $\frac{1}{8}$. No, ohhhh.
Student 1:	There's nothing under eighths.
Student 2:	This is ridiculous.
Student 5:	We need to get $\frac{2}{4}$ to get a whole. We need to get...
Student 6:	$\frac{5}{5}$.... $\frac{5}{6}$.
Student 5:	Next card. $\frac{3}{6}$.
Student 6:	$\frac{3}{6}$.
Teacher Voiceover:	One of the benefits to having students share their strategies and having students work together, is they bring things to their own language, they expand upon ideas with their examples and they question and challenge each other in a different way so that they can get to a truer meaning of the math.
Student 5:	Wait are we moving.... we are moving..
Student 6:	We're done. We can't move thirds because we already did.
Student 5:	Oh yeah. Ok so...
Student 6:	(picks a card) $\frac{5}{8}$.
Teacher:	Can you walk me through what you decided to do?
Student 7:	Ok, well, uh, we put it onto $\frac{2}{4}$ and then $\frac{3}{8}$. Joe was doing it

	with the blocks, though.
Teacher:	Do you want to explain to us how you used these?
Joe:	I took these and this was $\frac{7}{8}$
Teacher:	So you built $\frac{7}{8}$ first.
Joe:	And then you took fourths and put them there. We take these and move them up there.
Teacher:	What does this show you by putting them here?
Joe:	That that's one half.
Teacher:	It is one half. What else does it show you?
Joe:	It shows you $\frac{2}{4}$.
Teacher:	It shows you $\frac{2}{4}$. Why do you think I wanted you to look at that? What were we trying to figure out?
Student 7:	It showed like, we split up the $\frac{7}{8}$ and we made it into a different fraction. We made it into a $\frac{2}{4}$ and a $\frac{3}{8}$ and that kind of helped us break it up.
Teacher Voiceover:	One student decided to get the fraction pieces to help them figure out how to use the fractions that they were working with. They clearly at first didn't see to take $\frac{7}{8}$ and move it off of the eighths line, but once they were given the opportunity to explore it first they were able to figure out a way to move to another line or move in different ways.
Teacher:	What are some things we can do with $\frac{9}{10}$?
Joe:	What I was going to do was line it up.
Teacher:	Ok.
Joe:	And one more.
Teacher:	You need one more? I'll pull it out for you. Here's one more. What are these?
Joe:	Those are the sixths. I want to see if I can.
Teacher:	So you are trying to see if you can get any sixths?
Student 7:	Well its half so we could make $\frac{3}{6}$.
Teacher:	You can move on $\frac{3}{6}$ and
Student 7:	then on $\frac{4}{9}$.
Teacher:	Very good. OK, I need everybody to pause. Ok I'd like this time for you to share your strategies. We've played for a long time and I saw fabulous math thinking as I was going around. Louisa, do you have a strategy that you want to share?
Louisa:	First we had $\frac{1}{10}$ but then when we got $\frac{2}{10}$ we had to move it over to $\frac{3}{10}$ and then later on we had also gotten $\frac{9}{10}$ but when

	we were already at $\frac{3}{10}$ that would have been over the 10 line so when we had gotten $\frac{2}{10}$ we moved that up to the $\frac{1}{5}$ so the $\frac{10}{10}$ would be equal to 1 and we would have moved the fifths line because the fifths hadn't been anywhere yet.
Teacher:	Oh, so you backed up a redid steps so that you could make it work.
Louisa:	Yeah.
Teacher:	Very good. Dave, want to share?
Dave:	I picked a card and it had $\frac{2}{8}$ but I had a chip on $\frac{1}{4}$ so instead of going to $\frac{2}{8}$, $\frac{2}{8}$ is equal to $\frac{1}{4}$ so I just added $\frac{1}{4}$ and got to $\frac{1}{2}$.
Teacher:	Great. Why was he able to do that?
Student:	Because $\frac{2}{8}$ is equivalent to $\frac{1}{4}$ so instead of putting it down as $\frac{2}{8}$ he put it as $\frac{1}{4}$ and $\frac{1}{4}$ added to $\frac{1}{4}$ equals $\frac{2}{4}$.
Teacher:	Very good. Tamara, you want to share? Good.
Tamara:	First I added on $\frac{2}{6}$ and then I pulled off the card that said $\frac{6}{6}$. So instead of doing this all over I went 1,2,3,4 which is one whole and then I went up to half of six is three, that was four, so it went 5, 6 which got me to $\frac{1}{3}$ because I know that $\frac{1}{3}$ is $\frac{2}{6}$ so it would be $\frac{1}{2}$ of the third would be $\frac{1}{2}$ of the sixth.
Teacher:	Ok, can anybody try and rephrase what Tamara just said?
Student:	She was on $\frac{2}{6}$ so she pulled up the card $\frac{6}{6}$ and
Teacher Voiceover:	I found that through rephrasing kids are much more in tuned to the lesson. They know that they are going to need to rephrase. It builds in that accountability that yes there is something valuable here for you to be learning and now you need to be paying attention to that so that you can rephrase it. I also have kids rephrase it as an assessment for myself.
Teacher:	I think we are running out of time so we need to stop. What I would like to say is a couple things. First of all, I was really impressed with the number of strategies that you were able to use today. Tomorrow you are going to play this game again but we are going to see if the game is any different having to go to 2 and what more strategies you need to be thinking about tomorrow.
Teacher Voiceover:	My math skills have grown tremendously from teaching math differently to my students because I get right involved in the investigations with them and I have doubled what I knew. I have never considered myself this great mathematician and I still at points find, fractions, but as I work with my students and I see how you can manipulate the information very differently than standing there and just teaching, you know,

Magical Hopes: Manipulatives and the Reform of Math Education*

By Deborah Loewenberg Ball

This article begins with a story from my own teaching of third-grade mathematics.[1] It centers on an unusual idea about odd and even numbers that one of my students proposed.[2] The crux of the story, however, is the response I've received whenever I've shown a segment of videotape from that particular lesson to groups of educators.

First, what happened in the class: One day, as we began class, Sean announced, seemingly out of the blue, that he had been thinking that six could be both odd *and* even because it was made of "three twos." He drew the following on the board to demonstrate his point:

He explained that since three was an odd number, and there were three *groups,* this showed that six could be both even and odd. We had been working with even and odd numbers and exploring patterns that the children had noticed such as, "An even number plus an even number will always equal an even number." At this point, the definition of even numbers that we shared was that a number was even "if you can split it in half without having to use halves":

Six is even because you can split it in half without having to use halves.

Deborah Loewenberg Ball is associate professor of teacher education at Michigan State University in East Lansing. She conducts research on teaching and learning to teach mathematics and is especially interested in problems of changing practice.

*from the *American Educator,* Summer 1992. Magical Hopes: Manipulatives and the Reform of Math Education by Deborah Ball, 14–18, 46–47. Washington, DC: The American Federation of Teachers. Copyright ©1992 AFT. Reprinted with permission. All rights reserved.

Five is not even because you have to split one in half. Five is odd.

Sean was apparently dividing six into *groups of two* rather than into *two groups*. Although the other children were pretty sure that six could not be considered odd, they were intrigued. Mei thought she could explain what he was thinking. She tried:

> I think I know what he is saying . . . is that it's, see. I think what he's saying is that you have three groups of two. And three is an odd number so six can be an odd number and an even number.

Sean nodded in assent. Then Mei said she disagreed with him. "Can I show it on the board?" she asked. Pausing for a moment to decide what number to use, she drew ten circles and divided them into five groups of two:

Mei: Then why don't you call other numbers an odd number and an even number? What about ten? Why don't you call ten an even and an odd number?

Sean: *(paused, studying her drawing calmly and carefully)* I didn't think of it that way. Thank you for bringing it up, and I agree. I say ten can be odd or even.

Mei: *(with some agitation)* What about *other* numbers? Like, if you keep on going on like that and you say that *other* numbers are odd and even, maybe we'll end up with all numbers are odd and even! Then it won't make sense that *all* numbers should be odd and even, because if all numbers were odd and even, we wouldn't be even *having* this discussion!

I think this episode illustrates the dilemma faced by teachers who are committed to respecting students' ideas and yet also feel responsible for covering the curriculum. On the one hand, numbers are *not* conventionally considered both odd and even. Why not just tell Sean this and clarify for all the students that the definition of an even number does not depend on how many groups of two one can make? On the other hand, Sean *was* beginning to engage in a kind of activity that is essential to number theory: namely, noticing and exploring patterns with numbers, and, as such, his idea was worth encouraging. As the conversation unfolded in the class, Sean sparked the other children to discover that alternating even numbers (i.e., 2, 6, 10, 14, 18, etc.) had the same property he had first observed with six. Fourteen is seven groups of two, eighteen is nine groups of two, and so on. Each of these numbers is composed of an odd number of groups of two, and could be considered, according to Sean, both odd *and* even.

I have shown a small portion of the videotape from this class to other educators on several occasions. My intention has been to provoke some discussion about how to handle this situation: Should I seek other students' opinions? Clarify the definition of even numbers? Agree with Mei and move on to the plan for the day? Is this an opportunity or a problem to solve? Every time I show this tape, however, several teachers immediately inquire whether we used manipulatives for our work with even and odd numbers. When I say that we use drawings but did not use any concrete materials, these teachers have argued fiercely that that was "the problem" in this episode: Had I given the children counters as the medium for talking about even and odd numbers, then Sean would not have had this "confusion" about what makes a number even.

This response has baffled me. I am unable to discern how using counters and separating them into groups would have forestalled Sean's discovery that, if you group by twos, some numbers will yield an odd number of groups of two. Couldn't he have just moved six counters on his desk into three piles of two and made the same observation?

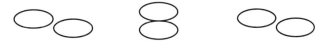

I am not convinced that manipulatives were the key to dealing with Sean's observation. Now, of course, I could have used manipulatives and told the children to divide the counters into two equal piles and if one were left over, then the number was odd. In other words, I could have guided their work more firmly toward the desired conclusions. But I could have done this in guiding their use of drawings as well. However, as a teacher, I am not necessarily interested in preventing the sorts of discoveries that Sean made. Moreover, I do not think that the point being made here had anything to do with whether the students were using manipulatives.

Some teachers are convinced that manipulatives would have been the way to prevent the students' "confusion" about odd and even numbers. This reaction makes sense in the current context of educational reform. In much of the talk about improving mathematics education, manipulatives have occupied a central place. Mathematics curricula are assessed by the extent to which manipulatives are used and how many "things" are provided to teachers who purchase the curriculum. Inservice workshops on manipulatives are offered, are usually popular, and well attended. Parents and teachers alike laud classrooms in which children use manipulatives, and Piaget is widely cited as having "shown" that young children need concrete experiences in order to learn. Some argue that all learning must proceed from the concrete to the abstract. "Concrete" is inherently good; "abstract" inherently not appropriate—at least at the beginning, at least for young learners. Whether termed "manipulatives," "concrete materials," or "concrete objects," physical materials are widely touted as crucial to the improvement of mathematics learning. From Unifix cubes, counters, and fraction pieces to base ten blocks, Cuisenaire rods, and dice, mathematics educators emphasize the role of manipulatives in promoting student learning.

One notable exception to this emphasis on manipulatives can be found in the *Professional Standards for Teaching Mathematics* (1991) published by the National Council of Teachers of Mathematics (NCTM). The use of manipulatives is not the centerpiece of this document's vision of mathematics teaching. Instead the *Standards* hold that teachers should encourage the use of a wide range of "tools" for exploring, representing, and communicating mathematical ideas. "Tools" include concrete models and materials, graphs and pictures, calculators and computers; and nonstandard and conventional notation. Manipulatives—or concrete objects—are important but no more so than other vehicles in NCTM's vision of mathematics teaching and learning. Still, because the passion for manipulatives runs so deep in the current discourse, many people read the *Standards* as a treatise that puts manipulatives at the center of mathematics teaching.

Manipulatives—and the underlying notion that understanding comes through the fingertips—have become part of educational dogma: Using them helps students; not using them hinders students. There is little open, principled debate about the purposes of using manipulatives and their appropriate role in helping students learn. Little discussion occurs about possible uses of different kinds of concrete materials with different students investigating a variety of mathematical content. Likewise, how to sort among alternatives, distinguishing the fruitful from the flat, receives little attention. Articles in teacher journals, workshops, and new curricula all illustrate how to use particular concrete materials—how to use fraction bars to help students find equivalent fractions, or beansticks to understand computation with regrouping. But rarely are alternative manipulatives compared side by side. For example, in teaching place value, what are the relative merits of base-ten blocks and beansticks? Is money an equivalently workable model? How do bundled Popsicle sticks fit with the other options available? Rarely is the relative merit—in a specific context—of symbolic, pictorial, and concrete approaches explored. In teaching fractions, for example, what is gained from using fraction bars? Might drawing one's own pictures offer other opportunities? And rarely is the difficult problem of helping students make connections among these materials examined. Many teachers have seen students operate competently with base-ten blocks in modeling and computing subtraction problems, only to fall back to the familiar "subtract-up" strategy when they move into the symbolic realm.[3] This lack of specific talk leaves teachers in the position of hearing that manipulatives are good, maybe even believing that manipulatives can be very helpful, but without adequate opportunities for developing their thinking about them as one of several useful pedagogical alternatives.

A close examination of some widely used instructional materials reveals an assumption that mathematical truths can be directly "seen" through the use of concrete objects: "Because the materials are real, and physically present before the child, they engage the child's senses . . . . Real materials . . . can be manipulated to illustrate the concept concretely, and can be experienced visually by the child" (p. xiv).[4] Teachers' guides also often convey the impression that, when students use manipulatives, they will most likely draw correct conclusions. This approach suggests that the desired conclusions reside palpably within the materials themselves.

One of the reasons that we as adults may overstate the power of concrete representations to deliver accurate mathematical messages is that we are "seeing" concepts that we already understand. That is, we who already have the conventional mathematical understandings *can* "see" correct ideas in the material representations. But for children who do not have the same mathematical understandings that we have, other things can reasonably be "seen":

—Can I have a few of the blue fraction bars—the thirds ones?" asks Jerome. Dina passes him two and he piles them with his other fraction bars. "Is four eighths greater than or less than four fourths?" asks Ms. Jackson. Jerome thinks this is a silly question. "Four eighths has to be more," he says to himself, "because eight is more than four." Lennie, sitting next to him, makes a picture:

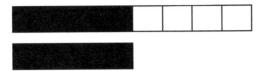

"Yup," says Jerome, looking at Lennie's drawing. "That's what I was thinking." But because he knows that he is supposed to show his answer in terms of fraction bars, Jerome lines up two fraction bars and is surprised by the result:

"Four fourths is more?" he wonders. He hears Ms. Jackson saying something about that four fourths means that the whole thing is shaded in, which is the same as what he has in front of him. It doesn't quite make sense, because the pieces in one bar are much bigger than the pieces in the other one. He does not quite understand what's wrong with Lennie's drawing, either. He moves some of the fraction bars around on his desk and waits for Ms. Jackson's next question. She asks, "Which is more—three thirds or five fifths?" Jerome moves two fraction bars in front of him and sees that both have all the pieces shaded. "Five fifths is more, though," he decides, "because there are more pieces."

Jerome is struggling to figure out what he should pay attention to about the fraction models—is it the number of pieces that are shaded? The size of the pieces that are shaded? How much of the bar is shaded? The length of the bar itself?

This vignette illustrates the fallacy of assuming that students will automatically draw the conclusions the teachers want simply by interacting with particular manipulatives. Because students may well see and do other things with the materials, some teachers strive to tightly structure students' use of manipulatives. This is usually done in one of two ways. One way is to use the materials that are relatively rigid. For example, if you use fraction bars to find equivalent fractions, it is difficult to come up with anything other than appropriate matches. The materials force you to get the right answers:

Find fractions that are equivalent to $\frac{1}{2}$

Magical Hopes: Manipulatives and the Reform of Math Education • 63

It is very hard to go wrong with these materials. Students' answers will likely be what we want: e.g., $\frac{4}{8}$, $\frac{2}{4}$ and so on. Another strategy often used to control students' thinking with manipulatives is to make rules about how to operate with the manipulatives so that students are less likely to wander into other conclusions or ideas. Fuson and Briars, for example, argue that any fruitful approach must lead the child to "construct the necessary meanings by using . . . a physical embodiment that can direct their attention to crucial meanings and help constrain their actions with the embodiments to those consistent with the mathematical features of the systems."[5] Nesher also emphasizes that any learning system must be built in with clear rules about how to use it.[6] For example, bundles of Popsicle sticks are often used to teach addition and subtraction with regrouping. Although the manipulatives in this case are relatively flexible, teachers will usually tell students that they must always group by tens and that when they need to subtract, they cannot do it unless they unbundle an entire group of ten. Without such instructions, many second graders I know would simply remove a few sticks from a bundle—just enough sticks to make the subtraction possible. But instead they follow the rules:

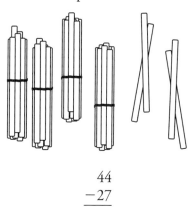

$$44 \\ -27$$

This works very well: Students unbundle a group of ten and count that they have fourteen sticks. Next they take away seven sticks. They then take two bundles of ten sticks away from the remaining three bundles, and they happily write down 17. Their answer is right. Following the rules, they readily arrive at the correct answers. In a sense, the manipulatives are employed as "training wheels" for students' mathematical thinking. However, most teachers have encountered directly the frustration when the training wheels are removed. Students, rather than riding their mathematical bicycles smoothly, fall off, reverting to "subtracting up" and other symbol-associated methods for subtraction. Even with close control over how students work in the concrete domain there are no assurances about the robustness of what they are learning. These training wheels do not work magic. Seeing students work well within the manipulative context can mislead—and later disappoint—teachers about what their students know.

My main concern about the enormous faith in the power of manipulatives, in their almost magical ability to enlighten, is that we will be misled into thinking that mathematical knowledge will automatically arise from their use. Would that it were so! Unfortunately, creating effective vehicles for learning mathematics requires more than just a catalog of promising manipulatives. The context in which any vehicle—concrete or pictorial—is used is as important as the material itself. By context, I mean the ways in which students work with the material, toward what purposes, with what kinds of talk and interaction. The creation of a shared learning context is a joint enterprise between teacher and students and evolves during the course of instruction. Developing this broader context is a crucial part of working with any manipulative. The manipulative itself cannot on its own carry the intended meanings and uses.

The need to develop these shared contexts was underscored for me when, in my class, we were using pattern blocks to develop some ideas about fractions. The children were able to build such patterns as:

and to label them as, respectively, two sixths and two thirds. They were able to interpret the two triangles as sixths in the first arrangement and the very same triangular pieces as thirds in the second. This attention to the unit is crucial both to understanding fractions in general as well as to using these blocks to develop such understandings. The students were also able to build arrangements that modeled other fractions, such as:

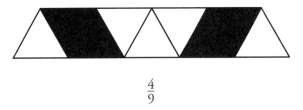

$$\frac{4}{9}$$

One day they were trying to figure out what one sixth plus one sixth would be. A disagreement developed between those who thought the answer was two sixths and those who thought it was two twelfths. Charlie argued that the answer had to be two twelfths, "because one plus one equals two, and six plus six is twelve."

$$\frac{1}{6} + \frac{1}{6} = \frac{2}{12}$$

Most of the children thought that made sense. Dalia disagreed and showed on the overhead with the transparent pattern blocks, that the answer had to be two sixths:

The other children easily agreed with Dalia. Following this, I thought the manipulative had convincingly helped students move toward the appropriate understanding until I heard Robbie explain, "*Both.* Both are right, because the answer is two twelfths with numbers, but two sixths with the blocks." Several others murmured assent. Juliette explained, "With numbers you add the one and the one and then you add the six and the six, and so you get two twelfths, but with the blocks, you have two of the one sixths, so you have two sixths." No one seemed at all disturbed that these answers did not correspond, and I realized that to know that these things were supposed to be congruent is something that has to be learned. The students had had plenty of experience with how context can affect both one's perspectives and one's answers. It made sense to them that the answers would vary in this case. They also had experience with mathematics problems having multiple solutions and, to them, this seemed like an example of such a problem. When Soo-Yung noted that Dalia's arrangement was also a picture of two twelfths (two pieces out of twelve), I knew we had a considerable way to go to use these materials toward some common understanding. Of course Soo-Yung was right. As was Dalia. I was beginning to understand how much work we needed to do in considering the question of unit in fractions.

The story of Soo-Yung and Dalia highlights the importance of the language we use around manipulatives. And how, even though they are more concrete than numbers floating on a page, there is much room for multiple interpretation and confusion. We need a lot more opportunity to discuss and develop ways to guide students' use of concrete materials in helping students learn mathematics. We need to listen more to what our students say and watch what they do. We cannot assume that apparently correct—or incorrect—answers, operations, or displays reflect the understandings that they appear to. Most of all, we need to put aside magical hopes for what manipulatives can do as we strive to improve mathematics teaching and learning.

If we pin our hopes for the improvement of mathematics education on manipulatives, I predict that we will be sadly let down. Manipulatives alone cannot—and should not—be expected to carry the burden of the many problems we face in improving mathematics education in this country. The vision of reform in mathematics teaching and learning encompasses not just questions of the materials we use but of the very curriculum we choose to teach, in what ways, to whom, and in what kinds of classroom environment and discourse. It centers on new notions about what counts as worthwhile mathematical knowledge. These issues are numerous and complex. For instance, we need to shift from an emphasis on computational proficiency to an emphasis on meaning and estimation, from an emphasis on individual practice to an emphasis on discussion and on ideas, reasoning, and solution strategies. We need to alter the balance of the elementary curriculum from a dominant fit focus on numbers and operations to a broader range of mathematical topics, such as probability and geometry. We need to shift from a cut-and-dried, right-answer orientation to one that supports and encourages multiple modes of representation, exploration, and expression. We need to increase the participation, enthusiasm, and success of a much wider range of students. Manipulatives undoubtedly have a role to play in these aims, by enhancing the modes of learning and communication available to our students. But simply getting manipulatives into every elementary classroom cannot possibly suffice to fulfill these aims.

Why not? First of all, much more support is needed to make possible the wise use of manipulatives. Many teachers, who themselves did not learn mathematics represented in a wide range of ways, do not find it easy to distinguish among a variety of models for mathematical ideas, nor to invent them for some ideas. Teaching with manipulatives is not just a matter of pedagogical strategy and technique. Few well-educated adults—not just teachers—can devise or use legitimate representations for many elementary mathematical concepts and procedures—from fractions to multiplication to chance. It should not be surprising to discover this. Consider merely the kinds of opportunities to explore and understand mathematics that most adults have had. Although a number are competent with procedures, many have not had the opportunity to develop the accompanying conceptual understandings that are necessary to manage the development of appropriate concrete contexts for learning mathematics and to respond to students' discoveries (e.g., Soo-Yung's observation that the arrangement of triangles on top of hexagons showed that $\frac{1}{6} + \frac{1}{6} = \frac{2}{12}$). Most adults simply remember learning that, with fractions, you do not add the bottom numbers. Why not? Few can explain or model it. And still fewer can explain what is going on with Soo-Yung's observation. Modeling addition and subtraction is one thing; modeling probability, factoring, or operations with fractions is another.

We also need to question and talk more openly about what we know about learning and about knowledge. Although kinesthetic experience can enhance perception and thinking, understanding does not travel through the fingertips and up the arm. And children also clearly learn from many other sources—even from highly verbal and abstract, imaginary contexts. Although concrete materials can offer students contexts and tools for making sense of the content, mathematical ideas really do not reside in cardboard and plastic materials.

More opportunities for talk and exchange—not just of techniques, but of students' thinking, of the pitfalls and advantages of alternative models, and of ways of assessing what students are learning—are needed. If manipulatives are to find their appropriate and fruitful place among the many possible improvements to mathematics education, there will have to be more opportunities for individual reflection and professional discourse. Like many other reforms, these sorts of support imply a need for restructuring. Delivering boxes of plastic link wooden cubes, and pattern blocks is insufficient to affect the practice of mathematics teaching and learning. At best, such deliveries can alter the surfaces of mathematics classrooms. They do not necessarily change the basic orientation to mathematical knowledge and to what counts as worth knowing. They do not necessarily provide students with conceptual understandings. They are not necessarily engaging for all students. In a few years the boxes of manipulatives will sadly be collecting dust in the corners of our classrooms, next to the artifacts of our past magical hopes. Manipulatives will continue to play a very important role—both as an appealing level to motivate and inspire change and as an important tool in teaching and learning. But it is time to stop pretending that they are magic and turn to more serious and sustained talk and work. Then we will begin to move beyond quick fixes and panaceas and face off with the difficult challenge of improving students' learning.

Acknowledgement

I would like to acknowledge several colleagues whose ideas about this article and about the uses of manipulatives have influenced my own thinking: Daniel Chazan, David Cohen, Magdalene Lampert, Dirck Roosevelt, Kai Suzuka, and Suzanne Wilson.

References

[1] I teach mathematics daily in a local elementary school in Sylvia Rundquist's third-grade class in East Lansing, Michigan. She and I have been collaborating since 1988; I teach mathematics and she teaches all the other subjects. In our regular meetings (and the conversation in between) we talk about the children, the culture of the classroom we are sharing, and about our role in helping students learn. My aim in this work is to investigate some of the issues that arise in trying to teach mathematics in the spirit of the current reforms (e.g., the NCTM *Standards* (1989, 1991). It is a kind of research into teaching that we see as complementary to other research on teaching.

[2] I have written about this story more extensively in "With an eye on the mathematical horizon: Dilemmas of teaching elementary school mathematics," which will appear in the *Elementary School Journal*.

[3] The "subtract-up" strategy, familiar to all elementary teachers, consists of looking at a problem like:

$$\begin{array}{r} 57 \\ -39 \\ \hline \end{array}$$

and computing $9 - 7$ instead of regrouping to subtract 9 from 17. This is one of the most persistent computational procedures that young children use.

[4] Baratta-Lorton, M. (1976). *Mathematics Their Way.* Menlo Park: Addison-Wesley.

[5] Fuson, K., & Briars, D. (1990). Using a base-ten blocks learning/teaching approach for first- and second-grade place value and multi-digit addition and subtraction. *Journal for Research In Mathematics Education,* 21, 180–206.

[6] Nesher, P. (1989). Microworlds in mathematical education: A pedagogical realism. In L B. Resnick (Ed.). *Knowing, learning and instrumentation: Essays in honor of Robert Glaser* (pp. 187–215). Hillsdale, N. Erlbaum., p. 188.

[7] In our research (e.g., Ball, 1990), we asked college students and other adults to make up a story, draw a picture, or use concrete objects to model division of fractions: $1\frac{3}{4} \div \frac{1}{2}$. Only a very small percentage of adults in any category were able to correctly represent this statement. Most modeled $1\frac{3}{4} \div 2$ instead of dividing by $\frac{1}{2}$. A sizable proportion said that this statement was not possible to model in any meaningful way.

SESSION 5

Supporting Generative Learning: How We Talk with Teachers

> Generativity refers to individuals' abilities to continue to add to their understanding . . . If teachers can learn to talk to their students about their thinking, puzzle about what the responses tell them about students' understanding, decide how to use this knowledge in planning instruction and interacting with students, and figure out how to learn more about the students' thinking—the teachers' own learning can become generative[*]

In this session you continue to develop an eye for mathematics classrooms using the Math Content Observation Guide. In addition, you begin to rethink how you talk with teachers about what you see. These two strands will continue throughout the remainder of the course.

As in the previous session, you will focus on how children's mathematical ideas develop gradually over time and how within any given class, children will be working on a range of ideas.

The first four sessions of this course have focused your attention on the importance of the mathematics being explored in the classroom. In addition to thinking about *what* you talk about with teachers, you will now think about *how* you talk to and with teachers. You will look at the modes of communication that you use when conferencing with teachers and consider how these modes of communication affect the post-observation conference.

In this session you will also explore the idea of "generative learning," the ability to continually add to one's understanding, and the supervisory practices that best support it. You will reflect on how generative learning might guide your work in supervising teachers. One useful idea that you will be asked to think about is that of "collaborative inquiry" (or "co-inquiry") in which teacher and administrator share their curiosity and their questions about the student thinking observed in the classroom, and collaboratively probe more deeply about that thinking. By adopting a stance of collaborative inquiry, you can express your curiosity about children's mathematical thinking and help teachers to become more reflective on students' mathematical thinking. This stance can help teachers become ongoing generative learners who continually adjust their instructional practice in relation to children's mathematical thinking.

---

[*]From Franke, M. L., Carpenter, T. P., Levi, L., & Fennema, E. (2201) Capturing teachers generative change: A follow-up study of professional development in mathematics. *American Educational Research Journal, 38*(3), 653–689.

Re-thinking How We Talk with Teachers and Observing for Content

A. What Is Generative Learning?

Read *Capturing Teachers' Generative Change: A Follow-up Study of Professional Development in Mathematics* by Megan Loef Franke, Thomas P. Carpenter, Linda Levi, and Elizabeth Fennema (Reading 7).

Skim pages 77–82, and then skip to page 87 and read more carefully.

Two notes:

- The idea of generative learning is expressed throughout this paper in several related terms: "generative growth," "generative change," and "generativity."

- This paper was written for a research audience rather than a practitioner one, so you may find it more technical than what you usually read.

Jot down some thoughts on the following questions:

1. How do the authors characterize generative learning?

2. What view of teachers-as-learners do the authors articulate?

3. What do teachers need to know to be generative learners?

4. Why does a focus on children's mathematical thinking help promote generative learning in teachers?

B. Observing for Content

Do an observation of one of the participating teachers in your school or district. Use the attached Pre-observation Conference Questions to help you prepare for the observation. Use the Math Content Observation Guide for these observations.

Write up a short summary of your observation, addressing as many of the questions in the Math Content Observation Guide as are relevant. Conclude with any questions or issues you are puzzled about.

C. Talk with Teachers

In the next class session we will focus on the "voices" we use in communicating with teachers about what we observe in their classrooms.

For this assignment you will need to have in mind a post-observation conference that you have conducted with a teacher. You can plan to conduct one with the teacher you have just observed or consider one that you did recently.

Reflect in writing on the kind of conversation that took place. Think about the nature of both your and the teacher's role in the conversation. Pay particular attention to the mode of communication that you employed. To what extent were you:

- giving advice?
- criticizing?
- asking questions and finding out what the teacher intended?
- sharing curiosity about students' thinking?
- problem-solving collaboratively on specific issues?
- praising or affirming?
- other?

Fill in the outline pie chart to show the proportions of each of these modes that you used in your end of the discussion. Be sure to use the color key on the chart. Copy your chart onto the large chart paper that the facilitator will give you.

Lenses on Learning, SESSION 5

Modes of Communication

	Color Key
Red:	Giving advice
Blue:	Criticizing
Yellow:	Asking questions and finding out what the teacher intended
Orange:	Sharing curiosity about students' thinking
Purple:	Collaborative problem-solving on specific issues
Green:	Praising or affirming
White:	Other

Pre-observation Conference Questions

In order to help you make sense of what you will be seeing when you do your classroom observations, plan to meet with the teacher prior to the observation and ask the following questions:

1. What topic will you and your students be working on in this lesson?

2. What do you plan to do in this lesson? (e.g., the origin and structure of the lesson, and so on)

3. What do you hope to accomplish in this lesson?

4. What mathematical ideas are embedded in this lesson?

5. What have you and your students been working on prior to this lesson?

6. How does this lesson fit into your overall goals for the year?

7. Are there students who have special issues in the class?

Lenses on Learning, Session 5

Math Content Observation Guide

	Students	
Focus Question	Conjectures	Evidence from Classroom
What mathematical ideas are embedded in the lesson?		
• What is the topic?		
• What are the ideas within this topic that are being explored?		
• What specific ideas are being explored by different students or groups of students?		
What makes this worthwhile mathematics?		
• What important mathematical ideas are involved?		
• What is the relationship between doing procedures and exploring ideas in this mathematics?		
• What kinds of mathematical thinking are taking place (conjectures, proofs, revising, generalizing, etc)?		

© Education Development Center, Inc.

Lenses on Learning, SESSION 5

Math Content Observation Guide

Teachers		
Evidence from Classroom	Conjectures	Focus Question
		What does the teacher seem to understand about the mathematics?
		• What aspects of the mathematics does the teacher seem to know well and in what areas does he or she still seem to need to deepen her knowledge?
		• How is the teacher responding to the mathematics in students' mathematical thinking?
		• How does the teacher show that he or she understands enough about the mathematics to test the boundaries of students' understanding?
		What does the teacher seem to understand about the development of children's ideas in this topic?
		• How does the teacher show that he or she understands students' thinking?
		• What follow-up questions does the teacher ask to probe the robustness of students' understanding?
		• In what ways does the teacher help them extend or deepen their thinking?
		What seems to be the teacher's long-term mathematical agenda?
		• What mathematical ideas is the teacher probing?
		• What does the teacher do to bring into focus the long-term importance of these mathematical ideas?

© Education Development Center, Inc.

Session 5 ◆ 75

READING 7

Capturing Teachers' Generative Change: A Follow-Up Study of Professional Development in Mathematics*

Megan Loef Franke
Thomas P. Carpenter, Linda Levi, and Elizabeth Fennema

This study documents how teachers who participated in a professional development program on understanding the development of students' mathematical thinking continued to implement the principles of the program 4 years after it ended. Twenty-two teachers participated in follow-up interviews and classroom observations. All 22 teachers maintained some use of children's thinking and 10 teachers continued learning in noticeable ways. The 10 teachers engaged in generative growth (a) viewed children's thinking as central, (b) possessed detailed knowledge about children's thinking, (c) discussed frameworks for characterizing the development of children's mathematical thinking, (d) perceived themselves as creating and elaborating their own knowledge about children's thinking, and (e) sought colleagues who also possessed knowledge about children's thinking for support. The follow-up revealed insights about generative growth, sustainability of changed practice and professional development.

MEGAN LOEF FRANKE is Associate Professor, Graduate School of Education and Information Studies, 1022 Moore Hall, University of California, Los Angeles, CA 90095-1521. Her specializations are teacher learning and professional development.

THOMAS P. CARPENTER is Professor, Teacher Education Building, 225 N. Mills St., University of Wisconsin, Madison, WI 53706-1795. His specializations are children's mathematical learning and the professional development of teachers.

LINDA LEVI is Associate Researcher, Wisconsin Center for Education Research, Room 575 D, Educational Sciences, University of Wisconsin, Madison, WI 53706. Her specializations are student and teacher mathematical learning and professional development.

ELIZABETH FENNEMA is Professor Emeritus, Room 575 G, Educational Sciences, University of Wisconsin, Madison, WI 53706. Her specializations are mathematics teaching and learning and gender differences in mathematics.

---

*from *American Educational Research Journal* 38 (3) Capturing Teachers' Generative Change: A Follow-Up Study of Professional Development in Mathematics, by M. Franke, T. Carpenter, L. Levi & E. Fennema. Copyright ©2001 by the American Educational Research Association. Reprinted with permission from the publisher. All rights reserved.

> I equate this with a journey. The children have been my teacher as well as people at the university and fellow teachers. . . . When you are on this journey there will be things that you as a teacher won't know, but you have to take a risk. You have to trust the kids. You do learn from the kids. You don't learn it in a week. You don't learn it in a workshop. You don't learn it in a month. You don't even team it in a couple of years. You continually improve. I think good teachers always are learners. . . . They have gotten hooked on how children learn and how they can best facilitate that learning, so they are always searching. That's what keeps teachers alive and vibrant, because they are always learning. You learn from kids, from fellow teachers, from the readings. You are always questioning. How can I be better? How can I be better, so my students are better? When you really start looking at kids, you see all the challenges. (Mazie Jenkins, teacher from Madison, Wisconsin)

Current reform ask teachers to make ambitious and complex changes. Such changes require more than being shown how to implement effective practices. Rather, as Little (1993) points out, teachers must engage in experimentation—discover and develop practices that embody central values and principles (p. 133), and become what Giroux (1988) calls teacher as intellectual. The reforms require that teachers reinvent their practices so that teaching and teaming are interdependent, not separate functions. "(Teachers) are problem posers and problem-solvers; they are researchers, and they are intellectuals engaged in unraveling the process both for themselves and for [their students]" (Lieberman & Miller, 1990, p. 112). Achieving this vision requires both changes in school culture as well as in individual teachers' conceptions of their practice and of themselves as learners. Achieving this vision also requires coming to understand what it means for a teacher to engage in ongoing learning, and then how professional development and the development of professional community can contribute to that end.

Sarason (1990, 1996) argues that a key to teachers' ongoing growth lies in creating school cultures where serious discussions of educational issues occur regularly, and where teachers' professional communities become productive places for teacher learning. Such schools, he argues, can replace teacher isolation with cultures of collaboration (Sarason, 1996). Indeed, studies of school reform demonstrate that powerful change can occur when the focus of the reform extends beyond the individual teacher (Hargreaves, 1994; Little, 1993). Collaboration among teachers seems to produce a greater willingness to take risks, learn from mistakes, and share successful strategies (Ashton & Webb, 1986). Just as classrooms promote student learning by becoming communities of learners where students collaborate to investigate topics in-depth, engage in collective reflection, and challenge each others' thinking (Cobb, Wood & Yackel, 1990), schools foster teacher learning when they become communities where teachers engage in challenging one another's thinking. The goal then is to create opportunities for teacher learning through professional communities whose activities are embedded in teachers' everyday work (see also Borko & Putman, 1996; Fullan, 1991; Lieberman & Miller, 1990; Secada & Adajian, 1997; McLaughlin & Talbert, 1993; Tharp & Gallimore, 1998).

Other scholarship attempts to understand teacher development by focusing on teachers as learners, examining what ongoing teacher learning looks like, how to accomplish it, and how to characterize the developmental trajectory leading to it

(For example, Borko & Livingston, 1989; Carpenter, Fennema, Peterson, Chiang & Loef, 1989; Leinhart & Greeno, 1986; Shulman, 1986; Simon & Schifter, 1991, Richardson, Anders, Tidwell, & Lloyd, 1991). Many of these studies report that when teachers reflect on their own classroom practices in principled ways, teachers can integrate their practical knowledge with research-based knowledge in ways that contribute to more successful practices (McLaughlin & Oberman, 1996; Schifter & Fosnot, 1993; Wood, Cobb, & Yackel, 1991). In some work this integration is termed practical inquiry (Franke, Fennema, Carpenter, Ansell, & Behrend, 1998; Richardson, 1990).

Work focusing on school communities is sometimes characterized as attending to the structural features of communities that support teacher learning, while work focusing on individual teachers addresses more specifically what teachers learn. However, as work on teacher learning evolves, this distinction is becoming less relevant. Researchers increasingly seek to characterize teacher learning, what leads to it, what supports it, and what it offers students, by understanding the relationship between structure and substance, and teachers and their communities. Peterson, McCarthey, and Elmore (1996), for example, point out that although school restructuring can provide opportunities for teacher learning, the organizational structures themselves do not cause learning to occur. They advise researchers to attend to the ways in which teachers create new ways to do established tasks within their classrooms and how the contexts in which teachers work influences how they acquire new knowledge and skill. The structure of the school community, how individual teachers learn, and what teachers learn all matter (Peterson, *et al.*, 1996; Richardson, 1990).

Learning with Understanding

Our work investigates how teachers, as members of reform communities, acquire knowledge that can become the basis for continued learning. The frame for our analysis of ongoing teacher learning draws on existing analyses of learning with understanding, Specifically, we focus on teachers' understanding of students' mathematical thinking and how it develops. One distinguishing characteristic of learning with understanding is that it is generative (Carpenter & Lehrer, 1999; Hiebert & Carpenter, 1992; Greeno, 1988). Generativity refers to individuals' abilities to continue to add to their understanding. When individuals learn with understanding, they can apply their knowledge to learn new topics and solve new and unfamiliar problems. When individuals do not learn with understanding, each new topic is learned as an isolated skill, and the skills they have learned can only be used to solve problems explicitly covered by instruction. A second defining characteristic of learning with understanding is that knowledge is rich in structure and connections. When knowledge is highly structured, new knowledge can be related to and incorporated into existing networks of knowledge, rather than simply added to existing knowledge incrementally, element by element. Consequently, learning with understanding is not only a matter of connecting new knowledge to existing knowledge, but also includes reorganizing knowledge to create rich integrated knowledge structures. A third factor in learning with understanding is that learners see learning as driven by their own inquiry. Carpenter and Lehrer (1999) propose that learners must perceive their knowledge as their own, believing that they can construct knowledge through their own activity.

In sum, knowledge becomes generative when the learner sees the need to integrate new knowledge with existing knowledge and continually reconsiders existing knowledge in light of the new knowledge. Building on the learning with understanding work, we view teachers' ongoing learning as learning with understanding. We focus on three particular features of teacher learning: (a) generativity, (b) creating structure, and (c) viewing knowledge about teaching and learning as constructed, self-created, and continually changing.

Focusing on Understanding Children's Thinking

Our view of teachers' learning with understanding presupposes that professional development with teachers must address what teachers learn. This positions our work along side a number of current professional development research projects that emphasize what is learned by teachers (Borko, Mayfield, Marion, Flexer, & Cumbo, 1997; Richardson, 1994a; Schifter & Fosnot, 1993; Sherin, 1997; Schifter & Simon, 1992; Simon & Schifter, 1991). Researchers within this field work to understand what would be helpful for teachers to learn. We are concerned with how what teachers learn can provide opportunities for teachers' ongoing learning. An approach that provides potential for ongoing learning focuses on teachers' learning about student thinking (Barnett & Sather, 1992; Brown & Campione, 1996; Dana, Campbell, & Lunetta, 1997; Lehrer & Schauble, 1998; Schifter, 1997; Schifter & Fosnot, 1993). Focusing on student thinking provides an opportunity for teachers' learning to become generative. If teachers can learn to talk to their students about their thinking, puzzle about what the responses tell them about students' understanding, decide how to use this knowledge in planning instruction and interacting with students, and figure out how to learn more about the students' thinking, then the teachers' own learning can become generative. Our goal, then, is to understand how teachers can make use of the opportunities provided within professional development focused on the development of children's' mathematical thinking to become engaged in generative change.

The study reported here examined how teachers understood and used knowledge of student thinking 4 years after their participation in a professional development program designed to help them construct frames for understanding the development of students' mathematical thinking. The purpose of the study was not simply to document whether teachers continued to implement principles of a particular program 4 years after the program ended; rather, it was an attempt to understand what characterized and supported teachers' continued learning and growth and what distinguishes teachers whose learning had become generative from those for whom this was not the case. We are particularly interested in the relationship between teachers ongoing learning and the professional communities developed in their schools.

Background

Cognitively Guided Instruction

The professional development program that served as the basis for this study is Cognitively Guided Instruction (CGI; Carpenter, Fennema, & Franke, 1996; Carpenter, Fennema, Petersin, & Carey, 1988; Fennema *et al.*, 1996; Franke, Fennema, & Carpenter, 1997). CGI focuses on helping teachers understand children's thinking by helping them construct models of the development of children's mathematical thinking in well-defined content domains. No instructional materials or specifications for practice are provided; rather teachers develop their own instructional materials and practices from watching and listening to their students and struggling to understand what they see and hear.

The first principle of CGI is that teachers develop a basis for understanding and building on their students' mathematical thinking. Corollaries of this principle are that teachers learn from listening to their students and struggling to understand what they hear. The knowledge about children's thinking that provides the basis for the program has inherent structure that can serve as a framework for organizing teachers' understanding. Within this structure, coherent principles relate the semantic structure of problems to the strategies that children use to solve them, and children's strategies evolve in predictable trajectories. The analysis of children's thinking that teachers engage in is robust. As teachers talk to students in their own classrooms, the teachers see how the analyses of children's thinking discussed in the workshops directly relates to the ways that students in their classroom think and learn. Ideally, this provides a basis for teachers to realize that they themselves can create knowledge about children's thinking as they interact with their own students and thus, in principle, CGI provides opportunities for teachers to continue learn with understanding.

Research on Sustainable Change

Traditional perspectives on teacher training concern themselves with the fidelity of teachers' practices to those specified in the training programs. The issues of long-term change focused on whether new practices could be sustained and remain faithful to the standard, not on the generation of new conceptions or practices. Examination of sustainability, in general, showed declines in retention over time. Current conceptions of professional development provide a different perspective on sustaining change and making it generative. Rather than training teachers to implement given practices, the interest is in having teachers come to see themselves as ongoing learners, seeking classroom practices that are responsive to the needs of the students and continually evaluating and adapting classroom practice.

Although professional development programs are becoming increasingly concerned with helping teachers develop a basis for ongoing change, few studies have systematically studied the long-term effects of these programs (Goldenberg & Gallimore, 1991). Most recent studies engaged in this type of investigation have focused on impact on teachers, students, and schools within a year of the professional development. A few studies have followed teachers more than a year after professional development in an attempt to understand sustainability and, at some level, generativity.

Richardson (1994a), for example, documented changes in 13 teachers' beliefs and practice 2 years after they participated in a professional development project emphasizing practical argument in teaching reading. They reported that the teachers sustained their practice, and in some ways, continued to develop and grow. Knapp and Peters on (1995) studied a cohort of first-grade teachers who had participated 3 or 4 years earlier in the initial CGI 4-week summer workshop. Nineteen of the 20 teachers interviewed stated that they still used elements of CGI in their classrooms, but teachers differed in how they conceptualized and implemented CGI. The teachers fell into three groups. One group reported that the knowledge of student thinking they developed in the CGI workshop had become the mainstay of their mathematics teaching. These teachers viewed CGI conceptually. Another group of teachers reported never having used CGI more than supplementally. They saw CGI as a group of procedures. The third group of teachers reported using CGI more at first but less currently; they evidenced an incongruity between their espoused beliefs and reported practices. Knapp and Peterson concluded that teachers' conceptualization of CGI influenced their conceptualization of the role of children's thinking in their classrooms. Knapp and Peterson's study illustrates that teachers' conceptions of professional development influence the degree to which teachers continue to Use CGI years later, yet it does not address directly issues of generative change as a function of professional development.

CGI longitudinal study. Since the initial CGI study upon which Knapp and Peterson's follow-up study was based, we have extended the scope of CGI and have conducted a 3-year longitudinal study of first-through third- grade teachers and their students (Fennema *et al.*, 1996; Franke, Fennema, Carpenter, Ansell, & Behrend, 1998).

Table 1

Timeline: Professional Development to Follow-up

Years of engagement		
1990–1993	1993	1997
Professional development, 26 first- through third-grade teachers, 5 schools	Final data collection, 26 first- through third-grade teachers, 6 schools	Follow-up data collection, 22 first- through fifth-grade teachers, 8 schools

We engaged with the teachers in professional development over a 3-year period and collected data over a 4-year period. During the first 2 years of the study teachers received regular support in the form of workshops, teacher meetings, and mentoring in their schools. During the third year, minimal support was provided, although researchers continued to visit classrooms on a regular basis and interview the teachers about their beliefs and practices. The study documented changes in teachers' beliefs and practice during the 3 years of the teacher development program. The current study examines the teachers 4 years after the professional development ended, and draws on both teachers' perceptions and our perceptions of how they have changed since their final 1993 interview. This study does not attempt to trace teachers' learning over time, rather it uses the initial 4 years of data as a base for understanding teachers 4 years later. We then examine the characteristics of teachers who had learned since the end of the professional development, both as individuals and in relation to the communities in which they participated. We present a comparative study of the teachers from the original longitudinal study to follow-up, a case study of 2 teachers, and an examination from the perspective of the teachers the supportive learning communities within teachers' schools.

Methods

Participants

We initially identified a sample of 26 teachers who participated in the CGI teacher development program from the spring of 1990 through the spring of 1993 (see Table 1). At the time of our current investigation, 3 of the teachers were no longer teaching mathematics and 1 teacher chose not to participate in the follow-up study. The 22 teachers in the final sample taught at six different schools in 1993 at the end of the longitudinal study (our final data collection point), In five of the schools almost all of the primary grade teachers had participated in the professional development program. The sixth school included 1 teacher who transferred from one of the other five schools. By the 1996–1997 school year, 2 teachers had transferred to schools where they were the only teachers who had participated in the professional development program. All other teachers remained in the schools they were at in 1993. In 1993, all teachers taught in grades 1–3. In 1996–1997, all teachers taught in grades 1–5: 6 teachers taught at a higher grade level than they had in 1993. Three teachers taught first grade; 7 teachers taught a first-and second-grade combination; 4 teachers taught second grade; 5 teachers taught third grade; 2 teachers taught a fourth and fifth grade combination; and 1 teacher taught fourth grade. Four of the teachers in this sample regularly taught CGI workshops to other teachers.

Procedures

Data collection occurred between November 1996 and May 1997. Each teacher was interviewed, and each teacher's mathematics class was observed with the exception of 2 teachers who were interviewed but did not consent to have their classrooms observed.[1]

Observations. All data were collected by one of the authors who had participated in data collection and analysis for the 1990–1993 longitudinal study and had demonstrated reliable classification of the levels of teacher development (see Fennema *et al.,* 1996). For each teacher, the classroom observer followed the teacher, taking notes on all teacher/student mathematical interactions. The observer focused on the problems the teacher posed, the strategies the students used to solve the problems, the teachers' interactions with students, and the interactions among students about their mathematical thinking. Of particular interest was how the teacher facilitated the students' sharing of their solution strategies; what the teacher did when students shared incomplete or unclear strategies; and whether or not the teacher elicited students' thinking beyond that required to solve the particular problems posed. The observer took detailed field notes and audiorecorded the class.[2]

---

[1] Both were classified at Level 2 (see following descriptions).

[2] Although we recognize that 1 day of observations may provide a limited view, we were looking for teachers' best (from their perspective) implementation of CGI, we asked teachers about the typical nature of the observed lesson, and we were interested in how the teachers discussed their practice what they knew and learned about their students' mathematical thinking.

Interviews. The teacher interviews occurred within 2 hours of each observation. The interviews were not scripted but were framed by a list of specific questions. The interviewer posed the initial question and then followed the teacher's lead, asking follow-up questions based on the teacher's responses. The interviewer cycled back to topics to elicit more detail. Each interview was audiotaped and transcribed.

The interviews centered on how teachers attended to and used children's thinking and how the teachers perceived their own change. The first set of questions paralleled those asked in the longitudinal study to provide a common comparison for characterizing teachers' level of engagement with children's mathematical thinking. The second set of questions focused on teachers' perceptions of their change and their feelings about the support that enabled their change.

The interview began by discussing the observed lesson. The interviewer asked why the teacher chose to pose the problems that she did and what she noticed about the strategies children used to solve the problems. The interviewer encouraged the teachers to talk specifically about what they knew about particular children and to describe how they used that knowledge to make instructional decisions. Follow-up questions focused on what the teacher knew about the strategies specific children employed during the observed class, why the teachers asked specific questions of certain children, and how they chose who would share their strategies with the class. The interviewer also asked more general questions about how the teachers decided what to teach, when to teach it, and how to teach it. The interviewer continued to cycle through these questions until she could classify a teacher in terms of the levels of teacher engagement with children's mathematical thinking (see following section for a description of the levels).

The second part of the interview assessed a teacher's perceptions of how and why she had changed since the teachers development program ended in 1993. Teachers were asked if their mathematics instruction had changed since the teacher development program had ended and, if so, how it had changed. Teachers who did not mention changes in their knowledge or use of children's thinking were specifically asked if their knowledge and use of children's mathematical thinking had changed since 1993 and, if so, what factors contributed to this change. Although we were interested in any factors a teacher thought to be critical to her change, each teacher was asked to specifically describe factors about CGI that influenced her change or lack of change. Teachers were asked to describe the type of support they currently had in terms of their mathematics instruction. If they did not specifically mention other teachers, they were asked if they regularly talked with other teachers about issues related to their mathematics instruction.

Analyses

Our analyses consisted of a series of passes through various aspects of the data. First, the data were parsed to be analyzed in terms of the comparison of the teachers' level of engagement with children's mathematical thinking, then all the data were examined for patterns related to generative growth. Following these analyses, we focused on cases of 2 teachers that exemplify the characteristics of generativity and the inter-relatedness of those characteristics. Finally, we examined the data to characterize the teachers' notions of support for implementation of CGI and their own learning more generally.

Levels of teacher development. To document teachers' engagement with children's mathematical thinking at follow-up and determine change from the end of the professional development program, we adapted a classification scheme used to characterize the teachers in the initial study (Fennema *et al.,* 1996; Franke, Fennema, Carpenter, Ansell, & Behrend, 1998). Our earlier scheme characterized the levels of CGI implementation. However, in our earlier work we separated beliefs and practice in characterizing teacher development (Fennema, 1996). In more recent analyses of the data, we found that because beliefs and practice are inter-related in complex ways, it is more productive to integrate beliefs and practice into a single scheme (Franke, Carpenter, & Fennema, 1997; Franke *et al.,* 1998). The levels of teacher development based on this scheme are summarized in Table 2. For more detail, see Fennema *et al.* (1996) and Franke *et al.* (1998), The scheme consists of a series of graduated benchmarks. It is not assumed that all teachers move through the benchmarks in a similar time frame, nor is it assumed that the process of learning is completely linear and unidirectional, The benchmarks indicate skills and understandings teachers have acquired and conceptions of how teachers think about the teaching and learning of mathematics. The benchmarks draw on the work of others who attempted to understand teacher learning and have been continually refined to capture most reliably the teacher's use of children's mathematical thinking.

Table 2

Levels of Engagement with Children's Mathematical Thinking

Level 1: The teacher does not believe that the students in his or her classroom can solve problems unless they have been taught how.
  Does not provide opportunities for solving problems.
  Does not ask the children how they solved problems.
  Does not use children's mathematical thinking in making Instructional decisions.

Level 2: A shift occurs as the teachers begin to view children as bringing mathematical knowledge to learning situations.
  Believes that children can solve problems without being explicitly taught a strategy.
  Talks about the value of a variety of solutions and expands the types of problems they use.
  Is inconsistent in beliefs and practices related to showing children flow to solve problems.
  Issues other than children's thinking drive the selection of problems and activities.

Level 3: The teacher believes it is beneficial for children to solve problems in their own ways because their own ways make more sense to them and the teachers want the children to understand what they are doing.
  Provides a variety of different problems for children to solve.
  Provides an opportunity for the children to discuss their solutions.
  Listens to the children talk about their thinking.

Level 4A: The teacher believes that children's mathematical thinking should determine the evolution of the curriculum and the ways in which the teachers individually interact with the students.
> Provides opportunities for children to solve problems and elicits their thinking.
> Describes in detail individual children's mathematical thinking.
> Uses knowledge of thinking of children as a group to make Instructional decisions.

Level 4B: The teacher knows how what an individual child knows fits in with how children's mathematical understanding develops.
> Creates opportunities to build on children's mathematical thinking.
> Describes in detail individual children's mathematical thinking.
> Uses what he or she learns about individual students' mathematical thinking to drive instruction.

The same author who observed and interviewed the teacher also coded teachers' level of engagement. In our earlier work, the author, along with her CGI colleagues, was trained to code level of engagement until reliability was reached across observers. Coding level of engagement involves reading holistically the observation field notes and the interview, looking for evidence that would support each benchmark and most importantly, looking for evidence indicating a teacher had not reached a particular benchmark. The coder must then make a written case for a particular benchmark, citing specific evidence from the data. If there were questions about a teacher's level of engagement, another author separately read the data and made a decision drawing on supporting evidence; if the coders disagreed, discussion revolved around the evidence and the group made a final decision.

Characterizing change. After coding teacher level of engagement, we reread all teacher interviews and observation notes to identify themes. Two authors (different from the author who coded level of engagement) read through all observations and interviews, keeping in mind our definition of understanding. We then discussed and elaborated the categories related to understanding, we characterized three categories: (a) specificity of talk about student thinking, (b) perceived use of frameworks or structures to organize knowledge of children's thinking, and (c) engagement in adapting and creating their knowledge about children's thinking. We came to consensus about how these categories could capture differences across teachers and whether they left out important teacher information. In our discussion we found that the characteristics developed from our definition did not include information on the degree to which teachers see student thinking as a central aspect of their practice. We added this category to the others: (d) centrality of student thinking.[3] So we took these four categories and systematically, cycled through the data.

---

[3] We coded the data by looking to see the degree to which our definition of understanding differentiated teachers. Thus, there may be characteristics that exist relevant to ongoing learning that we did not explore. We wanted to examine the robustness of the characteristics hypothesized by the theory.

One of the authors coded each interview for instances of each category. Rather than code each interview for all categories, the author coded each category across all interviews and then moved on to the next category. Each instance of each category was coded. Another author coded the interviews for confirming and disconfirming evidence in each category, specifically identifying disconfirming evidence. After completing the coding, the data were summarized by pulling out the relevant evidence from each interview by category. We could examine all instances of each category coded for each teacher and then examine all instances across teachers.

During the last coding phase, a sample of the data was coded for reliability. Three graduate students, knowledgeable about the development of children's thinking but not involved in the project, coded the data by category. Their coding was used to substantiate or identify conflicts in the original coding. We could compare if all instances of a category were captured and we could compare to see if all instances were coded within the same categories across coders. When conflicts were identified they were discussed and consensus was reached about recoding or refining the categories.

Case exemplars. The case exemplars were chosen to highlight one of the more subtle differences in the data related to generativity. The two case examples provide an illustration of how the characteristics of generativity work together in providing a picture of the teachers ongoing learning. The two teachers articulated different notions about the knowledge of children's mathematical thinking and how that can be used to drive practice. In documenting the cases we pulled out the coded episodes for each category for each teacher, synthesized the episodes, and compared them to the details of the classroom observation.

Support. We began our analyses of teachers' perceived support for CGI by pulling out, by teacher, each remark about support. We examined both what the teacher said about the form of support and the content of the support (Hargreaves, 1994). We synthesized the episodes and looked for patterns. Several patterns carried across teachers' conversation, however, we recognized that the form and content of support described were related to the school context. We describe the data patterns that we found within schools and then examine those patterns across schools to better understand the salient issues for teachers.

Results

Overall Teacher Change

All 22 teachers maintained some level of CGI implementation, yet some change in levels occurred from the end of the professional development program to follow-up (Table 3). All teachers who ended the professional development program at a high level of engagement with student thinking continued to exhibit high levels of engagement. Nine of the teachers who were at Level 4B at the end of the professional development program remained at Level 4B; 1 teacher went from Level 4B to a Level 4A. The 1 teacher at a Level 4A remained at a Level 4A at follow-up. The teachers at Level 3 were less consistent. Of the 9 teachers who ended the professional development program at Level 3, 4 remained at a Level 3, 1 moved to a Level 4B, and 4 moved to a Level 2. The 2 teachers who ended the professional development program at a Level 2 remained at a Level 2 at follow up. As the discussion below documents, these teachers continued to grow and evolve.

Table 3

Shifts In Teachers' Engagement with Children's Thinking from Project End to Follow-Up

	Teacher level at follow-up				
	1	2	3	4A	4B
Teacher level at project end					
1	0	0	0	0	0
2	0	2	0	0	0
3	0	4	4	0	1
4A	0	0	0	1	0
4B	0	0	0	1	9

Characteristics of Self-Sustaining Generative Change

Coding the teacher data using the levels of development criteria developed for the professional development project provided one basis for examining teacher change. This pass at the observation and interview data allowed us to characterize the extent to which teachers maintained their beliefs and practice from the end of the longitudinal study to follow-up. However, we found through our analysis that the data from the follow-up study also offered a new perspective on the levels, particularly with respect to self-sustaining, generative change. The existing levels were developed outside the context of generative change and focused on identifying the extent to which teachers made use of the development of children's mathematical thinking in constructing their beliefs and practice. Our new data provide insight into characteristics that allowed teachers to not only sustain particular practices but also continue to grow. In the next sections we revisit the teachers' data and describe how the four characteristics serve to enhance the teacher levels and contribute to our understanding of generative change.

Children's Mathematical Thinking as Central

When teachers described their thinking about the teaching and learning of mathematics and how that played out in their classrooms, they often described what they wanted their students to learn, what their students already knew, and what the curriculum said children in their grade should be learning. The degree to which children's thinking informed these issues clearly differentiated the Level 2 teachers from the Level 3, 4A, and 4B teachers. Although the Level 2 teachers felt that they considered children's mathematical thinking in their decision making, they reported that children's thinking was only one of an assortment of programs, issues, and approaches they considered. The Level 3, 4A, and 4B teachers, on the other hand, viewed children's thinking as driving every aspect of their mathematics classroom practice and described both its specific utility in the mathematics classroom and the more general utility of focusing on student thinking for other content areas.

For the Level 3, 4A, and 4B teachers, children's mathematical thinking was a constant. They integrated children's thinking into all that they did in mathematics. They spoke about children's thinking as something they, as teachers, could never give up; it was a part of who they were as teachers.

Ms. Nathan, a Level 3 teacher, pointed out how children's thinking, although not the only thing she did, was a part of her. "I think now it (children's thinking) is more a part of me, whereas in the beginning you weren't sure exactly what you should do or what you shouldn't do. . . . So l think now it's more natural."

Ms. Rothman, a Level 3 teacher, concurred,

> I'll first of all go back to the CGI training I had. And I think I said back then that I'd never go back to where I was, which was maybe like straight computation sort of thinking and drill. Because of that background, I think that's always there in my mind, no matter what I'm doing. And it's been tremendously helpful. In organizing my thoughts and giving story problems, in numbers that I might give, in all that, that's always there.

Each of the teachers at the higher levels commented that children's thinking drove their practice. They each provided explicit examples of how children's thinking drove their interactions with parents, their principal, other teachers, and their students.

Centrality of children's thinking clearly differentiated the Level 3 teachers from the Level 2 teachers. The Level 2 teachers did not view children's thinking as central. The Level 2 teachers talked throughout their interviews about issues of time. They felt that they had so much to do and cover that there was not always enough time to do justice to children's thinking.

> We do problems like this (word problem) and I use the math book but we do this type of problem on Friday when we do our superstar math or our problems of the day. . . . The thing that (is) really, really hard, I don't know whether this can ever be solved, but I feel like it's always such a fight to get things (to) fit into the day. I'm always fighting with everything. How can we have enough time to do this or how can we have enough time to do that? It's just a constant thing. (Ms. Michael, Level 2)

They told us that they could not always make decisions based on children's thinking because there were so many other issues to take into account, like content coverage, grade level expectations, the child's self-esteem, classroom management, and parent concerns.

Frameworks

Just as the centrality of children's mathematical thinking differentiated the Level 2 teachers from the teachers at the higher levels, we found that the degree to which teachers saw their knowledge of children's thinking as organized or structured differentiated the Level 3 teachers from the Level 4A and 4B teachers. A major tenet of CGI has been that teachers can structure their knowledge of children's thinking in ways that allow them to understand the principles supporting it. That is, the analysis of the mathematics of children's thinking involves more than lists of problems and strategies; and thus it involves an integrated perspective based on relationships among problems and strategies. Although the Level 3 teachers saw children's thinking as central, they did not focus on the structure of the knowledge of children's mathematical thinking.

The Level 4A and 4B teachers talked about the development of children's thinking in terms of a framework or structure. They saw the relationships among the problem types and strategies as critical to understanding what they would do with the knowledge.

> I need a framework. I definitely think there's a framework with CGI that's made a big difference for me. Strategies have been identified, there's definitely a hierarchy. That's helped. . . . I mean, a lot of curriculum materials have problem solving. A lot of them do. But you don't know what to do with it. I mean, how do you decide why problems would be more difficult than others for children to solve? You know, what makes this problem difficult? And with CGI, that has been researched, and I think accurately researched, and it enables me to know why certain kids are struggling, what I can do to facilitate that. So I think a framework is critical. For a person like me, especially. . . . I want to know where I'm going and where I'm taking them. So I need that framework. (Ms. Sage, Level 4B)

Each Level 4A and 4B teacher talked about frameworks or structures. In all cases the framework was a way of organizing their knowledge of children's thinking so they would have some sense of how to make decisions. One teacher talked about it as having "slots" that helped her think about where her students were and what problems to pose next. Another teacher talked about the structure as providing general guidance that she could then build on and fill in the details.

The Level 3 teachers saw the knowledge about children's thinking as a set of problem types that enabled them to think about problem difficulty, and a set of strategies that enabled them to see what children in their classrooms might be doing. They also talked at times about the knowledge being organized, but the organization related to either the problems or the strategies separately.

> I think what was revolutionary to me, at the very beginning of CGI, maybe scary to begin with, was that the story problems can be stated different ways . . . And I hadn't really thought through that. . . . I hadn't probably thought a lot about different kinds of strategies . . . So it's helped me to see children in a different light (Ms. Rothman)

This was as much detail as Ms. Rothman provided about how she thought about the knowledge of children's mathematical thinking discussed in the CGI workshops, The other Level 3 teachers spoke primarily about the problem types. They talked about how the problem types helped them know something more about what problems were possible to pose and how knowing something about problem difficulty enabled them to make instructional decisions. They talked about learning that children could invent strategies for solving these problems, but none of the Level 3 teachers ever talked about frameworks for organizing their knowledge of children's thinking.

Specifics of Student Thinking

As the Level 4A and 4B teachers talked about the structure of their knowledge, they also described how they saw the structure as enabling them to learn the details of children's mathematical thinking. When we examined the specificity at which the teachers could describe their children's mathematical thinking, we found that this also differentiated the Level 3 teachers from the Level 4A and 4B teachers. The Level 4A and 4B teachers articulated step-by-step descriptions of the strategies a student used to solve a given problem, and they routinely talked about the understanding underlying the student's solution. The Level 3 teachers described student thinking in general terms; when they were specific about a child's thinking they would talk about the types of strategies the child used.

Every Level 4A and 4B teacher detailed multiple children's thinking in their interviews with little or no prompting by the interviewer. In each of these cases we could tell from the teachers' descriptions exactly how a given child had solved the problem. This focus on detailing children's thinking was also apparent in the teachers' classroom interactions. In each Level 4A and 4B teachers classroom observation, the teachers consistently probed student thinking. Here the teachers would go beyond a single, "Can you tell me what you were thinking?" and if necessary, ask the student three or four times to tell them more about it. At times these teachers would also ask specific questions, such as, "Can you tell me how you got the 27?" or "I didn't see what you did with the cubes for this part."

The specificity at which the teachers knew children's mathematical thinking also extended to their explanations of the development of children's thinking, particularly the details of how students progress in their understanding. These descriptions emerged most often as teachers discussed why they had chosen the problems for the children.

> And I'm really seeing a lot of kids so far who are pretty much following the action of the problem. And so when I give problems that do not have action I'm more apt to get strategies where they might add and they might subtract. And I'm trying to get them to think about that as far as applying it to problems with action. In some cases it would make total sense on a join problem to subtract, just go two numbers back, and they really follow the action. . . . You know, because many of them think that they have to directly model that that's the only strategy that they have. I'm finally starting to see some counting strategies. Well, with the numbers 49 and 101, 1 really was hoping to see some of the kids use some of their knowledge to help them. . . . So that was the reason why I chose that. The multiplication problem, I chose the numbers because I did want a lot of them to use 10. But I'm also focusing on just getting the concept of multiplication down. I really see mostly direct modeling. I'm starting to see some kids do some skip-count, but just one or two numbers is about it. So that's the reason why I chose that. (Ms. Sullivan)

The Level 3 teachers focused on children's abilities to solve problems in a variety of ways. They valued the children's solutions, not in terms of the specific strategy the child used but rather in terms of having the children use and share different strategies. The lack of probing we observed in these teachers' interactions with students carried over into the generality in how they described student thinking in the interviews. Often, Level 3 teachers could not explain their students' thinking. At times they told us that they were not sure what a given student had done; at other times they made general inferences about why a child had difficulty with a problem that they could not support with specific detail. For example, when Ms. Mason was questioned about her observation of one student, she began by describing the student in general terms as "what I would call a slow learner." Then she discussed the tool the child used to solve the problem, a hundreds chart. Ms. Mason never described the specific strategy the child used to solve the problem with the hundreds chart. Each teacher at Level 3 and 2 was more likely to talk about the fact that a child used his or her fingers or did it in his or her head than to describe a specific strategy. When asked, teachers at Level 3 could describe strategies they had observed, but they did not do so when they were asked in their interviews to describe their children's thinking or explain the decisions they had made.

Coming to Know

Teachers at Level 4A and 4B focused on structure and detail in talking about their knowledge of children's mathematical thinking. The teachers differed in how they viewed this knowledge. Teachers at Level 4B saw the knowledge of children's thinking as their own. These teachers did not see the knowledge as research knowledge, or the professional development program leaders' knowledge. This was knowledge that they could create and adapt. Level 4B teachers saw themselves as constantly testing their knowledge, learning from their students, as engaging in practical inquiry. The teachers at Level 4A did not share this perspective.

Engaging in practical inquiry focused on the development of children's mathematical thinking was what distinguished the teachers at Level 4B. The teachers at this level were consistently curious about how their children would solve problems; they regularly tested and revised their knowledge. They thought about what the students did, and they struggled to make sense of it. They talked about themselves as learners. They described their learning processes and talked in detail about what they were learning. Every Level 4B teacher talked about how much they learned with each interaction they had with a child and how that knowledge could inform other interactions.

> I'm always learning from the kids. I mean, in terms of all the different ways to solve problems. All the different kinds of things you can ask kids to do and some of this I found this year because almost all the kids who are in second grade in my room this year I had last year as first graders, And it is true, because I knew them that much more or I feel that much more comfortable. (Ms. Karl)

> And so every time you interact with a child, you're gaining more knowledge of how to interact with other children. Every time they show you, and tell you, what they're doing and thinking, you just learn more about what's going on in their head. (Ms. Baker)

> "I think that I, it's kind of like I'm trying something and then I'm testing it and then I'm refining it, and then I am going again." (Ms. Sage)

Level 4B teachers reported that the substance of what they knew about children's thinking had changed.

> I think the point is that very often a kid isn't all one thing or another and there's some . . . not only do they flip sometimes from, oh I'm going to say direct modeling to counting, but then there's that kind of in between thing where they will, for instance, take cubes, take some kind of manipulative or tally marks or something or another and make both sets. But then maybe ignore the first set or just look at it and say, well, that they are going to add it. That is 7 and then counts up on the other one. Well do you call that modeling? Do you call that counting? Do you call that they're almost there? Well I call it they are almost there. And so that Is one of the places where it (the framework) does not quite, I mean that is one whole spot that isn't on the chart. . . . We can call it transition and go from there. (Ms. Karl)

The CGI teacher development program was based on models of children's thinking that we thought would be accessible and easy for teachers to use, The Level 4B teachers took those frameworks as a starting place to generate and test their own ideas about children's thinking. They elaborated those frameworks and made them more detailed. Sometimes Level 4B teachers reported that they thought about an aspect of children's thinking differently than they had thought about it before, and sometimes they reported that they disagreed with the distinctions discussed in the professional development program. Ms. Castelbury described how she had added to the frameworks that she had learned in the professional development program by closely examining her students' work. In the readings for the teachers and in the workshops themselves the relationships between problems and strategies had been summarized by what Ms. Castelbury referred to as the bubble chart.

> Well, I mean I don't think we really have a (a set of levels of development for all the content areas) well, for instance, well the bubble chart. There's no division on the bubble chart. There's no multiplication on the bubble chart. But still I'm working through those levels of modeling, counting, whatever. The same thing with fractions, the same thing with decimals, it's kind of still, there's sort of this overriding thing of, if you do their work and you collect it and you spread it out you, you kind of group it.

Another Level 4B teacher pointed out that there was always more to learn, yet for her learning was not just accumulating more details, but enhancing the knowledge that she already had about children's thinking.

> Because if you have an understanding of those basic ways kids approach problem solving, I think it just helps you . . . you know what to look for with them. But they don't always do it exactly that way. We have some strategies that kids use, but every time I go to fill out a (bubble chart), or take notes on the child, I go, wait a minute, let's put that here. . . . You're going to see different ways, but that's kind of your basic core knowledge. From that you keep adding. I think we could probably expand that. If we all got together, we could probably expand that whole flow sheet and make it humongous with different ways the kids have approached these problems. Because there's just not any one way that you can always expect them to do it. (Ms. Baker)

The teachers at Level 4A and lower did not discuss learning more about children's thinking on their own. These teachers articulated that they were learning (at least in some way), that they saw themselves differently, but they were not developing deeper understanding about children's thinking. When probed more specifically about what she had learned about children's thinking since the end of the project one Level 4A teacher talked about feeling more comfortable with the knowledge of children's thinking.

> I don't walk around with a clipboard any more, with the definitions of what these things are, like I used to, but I think you see kids, their learning styles and that the big thing about how kids learn. That has helped me accept individual differences and thinking that it's okay if someone knows derived facts, but you also have to show me, you also have to model, so that type of thing. . . . So it's part of my style now. . . . I think you laid the foundation for me, but it's just become part of my style or what I look for in kids now. (Ms. Park)

Another Level 4A teacher articulated how she consistently saw new things from her students but concluded that her basic knowledge had not changed.

> We're always seeing things that the kids are doing. I mean, sometimes you think you've seen it all, and then you see something wild and different. Anyway, as far as the story types, no, that doesn't change for me. Like I said, you might adapt a problem or something, but you pretty much stick with those basic story types. And as far as explaining what the kids are doing, we use a lot of the same terminology. So I'd say we kind of stuck with that basis that you gave us. At least I do. (Ms. Andrew)

A Level 3 teacher talked about learning about children's thinking from listening to what was taught in the workshops. "Maybe I didn't listen close enough when we had the class with regards to, then where to give them the next problem, to progress them. Maybe that's a weakness of mine." (Ms. Mason)

When asked specifically to describe how they had changed, the teachers at Levels 4A, 3, and 2 reported that they knew much the same types of things about children's thinking as they had when the teacher development program ended.

> No, it (my knowledge) hasn't changed. I probably have learned more, because you learn more as you do with kids. So I've learned more. I can't say that the knowledge, the basic fundamental knowledge that I learned, has changed, but I am changing in the sense that I'm learning more about children, and being more aware of children. I still keep my clipboard. I just keep it with me. Now we're going to be going to do our report cards. It's so easy for me to just go and take a look at that. And I have all the different kinds of story problems, and the child's name, and I just put my little things in there and I know, by these little symbols, what I'm talking about. (Ms. Rothman, Level 3)

Ms. Rothman's response was typical of the responses we heard from the Level 4A, 3, and 2 teachers. Ms. Rothman did not feel that her knowledge had changed, but she did think that she had been learning. Her learning is harder to pinpoint but she seemed to see it as learning more about individual children, not about creating a deeper understanding about how children's thinking develops.

Teachers at Level 3 and 4A thought that the knowledge about children's thinking was critical and central to their teaching, but they saw the knowledge as something passed on to them. Ms. Mason stated that more inservice might help her figure out how to use her knowledge of children's thinking. She wanted to learn how the researchers thought she should be using children's mathematical thinking, essentially, how she could do it right.

> And its [CGI] ingrained in me. I don't think it's ever going away. It's taught me to watch kids. I feet like I'm a totally different math teacher than I was, and I like that feeling, but I just don't think I'm doing it right yet. I don't know why. (Ms. Mason)

Integrating Characteristics of Generative Change

The characteristics of generative change are not a set of separate characteristics, but rather an interactive set that fit together in ways that allow for a better understanding of ongoing teacher learning. Here we provide the cases of 2 teachers that detail how the characteristics fit together to create a picture of teacher learning. The cases we have chosen, Ms. Sullivan and Ms. Carroll, highlight how the characteristics of generative change act together to differentiate a Level 3 from a Level 4B teacher. We chose teachers at these levels because we have seen the Level 3 teachers sustaining their practice and the Level 4B teachers as becoming generative in their growth. Examining the differences in Ms. Carroll and Ms. Sullivan's responses to the characteristics of generative change provides an opportunity to understand ongoing learning within the context of classroom and school communities.[4]

Ms. Sullivan

Ms. Sullivan clearly believed that children construct mathematical knowledge. She described the teacher's role in teaching mathematics as helping students build more understanding. When asked how she built understanding, she responded, "I don't think that I build it. I think they build their understanding." Ms. Sullivan held corresponding notions about her own knowledge development. She did not believe that her knowledge of student thinking was simply a product of the CGI professional development program, but rather as something that she continually reconstructed. She described how her own knowledge had grown and continued to grow by engaging in practical inquiry during her daily interactions with her students. When asked why she thought her knowledge had grown, Ms. Sullivan responded:

> Because I keep watching the kids, I learn from the kids a lot. I really do. I even stop and think a lot more conscientiously about my numbers. Even

---

[4] The comparison of the two cases is not intended to represent a contrast of expert and novice teachers. The authors share a great deal of respect for both teachers. In fact we would be happy to have our own children in either of their classes. Comparisons are difficult, because they seem to imply value judgements. That is not our goal. Both teachers were successful teachers who taught in ways that were consistent with current reform recommendations. The two classes had a great deal in common and provided students very similar learning opportunities. Both teachers respected their students and valued their intuitive understandings of mathematics. They both encouraged students to construct their own strategies for solving problems, and both provided opportunity for students to share and discuss alternative solutions.

thinking, if I pick [particular] numbers, what are all the possibilities that could happen.

I think [my knowledge of children's thinking] can't help but change, because I think I am more able to understand and break things apart, and build on kids' knowledge the longer I use this and see it developmentally. . . . I think definitely you cannot help but grow, because there is still more knowledge out there to be explored, and so I keep looking for that.

[I] haven't changed as far as the philosophy or my style of teaching. [I've] changed as far as my understanding of how third-graders think, and how third-graders would go about solving mathematical problems, and so like learning their levels of development and the strategies they would use, that's definitely changed. I've found myself more comfortable and really understanding them . . . more knowledgeably about the numbers to pick, especially when choosing numbers that would lend themselves to developing their place value concepts (and) invented strategies.

Ms. Sullivan not only spoke in general terms about how her knowledge had changed; she also described specific changes in her fundamental understanding of children's thinking. She knew a great deal about her individual students but, more critically for her continued growth, she used this specific knowledge to elaborate and extend the conceptual frames she used to understand children's thinking. As a consequence, she had much more detailed and complex schemes for analyzing children's thinking than the ones that were discussed during the professional development program. The workshops and readings had focused on distinctions between several primary classes of strategies that children invented for adding multi-digit numbers. Ms. Sullivan not only identified more fine-grained distinctions among strategies, she integrated them into a coherent framework that provided for her a more complete picture of the development of children's mathematical thinking.

I'm definitely more aware of how (the children) start out at the beginning of the year with modeling and how that modeling very much influences the types of invented strategies that they're going to use. . . . I'll see some kids count all the tens, and then they'll go to the ones, and then I'll see other kids who do 46, 56, 66, 76, and then pick up the ones. So for me there is a little bit of a distinction there that I want to clarify because it helps me know what they're probably going to do with it next. Do they do all the tens and then go to the ones? Or do they do a mixture? Because I really will see more kids use a tens and ones invented strategy when they do that versus more of an incremental strategy when they do the other one.

Using knowledge of student thinking. Ms. Sullivan's constructivist perspective both reflects and is reflected in her knowledge of her own students and in her use of that knowledge in planning and implementing instruction. Ms. Sullivan constructed more elaborate frames for analyzing students' mathematical thinking as she learned about her own students' thinking. The richer conceptual frameworks allowed her to acquire deeper and more detailed understanding of her own students' thinking. In the interview, Ms. Sullivan demonstrated specific knowledge of the strategies her students would use to solve different problems, and she clearly articulated how she based her instructional decisions on this knowledge. She described how she used what she knew about students' mathematical thinking to select problems and to decide how to interact with them about their solutions. She not only demonstrated specific knowledge of the students in her class in terms of the specific kinds of strategies they could use, but she also had a good map of where they stood in terms of developing more sophisticated strategies. She had well reasoned hypotheses about the strategies that might fall in the zone of proximal development for specific students and how problems might be structured to encourage those students to use those more sophisticated strategies. For example, she reported that because many of the students in her class relied on directly modeling the action in problems, even when other strategies might be easier and more efficient, she posed problems without specific action to encourage the students to realize that they could use a greater variety of strategies for the same problem. One of the problems she picked for the class we observed involved comparing two quantities and the students did use a variety of strategies to solve the problem.

Ms. Sullivan's choice of numbers also was based on her specific knowledge of children and her expectations of how their current strategies could be extended. The numbers she selected for the compare problem were 101 and 49. She had observed that certain children could solve problems by counting by 10, but only when they could count by decade numbers (50, 60, 70). By using numbers that were close to decade numbers, Ms. Sullivan anticipated that several students might realize that they could add 1 to the 49 to get 50 and then count up to 100 by tens or simply recognize that 100 is 50 more than 50. In discussing why she selected the problem and what students actually did, Ms. Sullivan talked about the strategies of specific children. She described how 1 child did use the target strategy counting by ten from 50 whereas another child counted by 10 from 49 (59, 69, . .). She also talked about the difficulty another child had in trying to figure out how to deal with the 1 that he added to 49. Another problem given the same day involved multiplying 9×9. Ms. Sullivan reported being curious about whether certain children would see that they could solve it by counting by 10 and then compensating by subtracting or counting back 9 from 90.

Classroom interaction. Ms. Sullivan's classroom practices were consistent with her goals of providing her students the opportunity to construct deeper understandings of mathematics and herself the opportunity to construct deeper understanding of students' mathematical thinking. During the class observation, Ms. Sullivan posed two problems that children solved individually. After each problem was solved, the class discussed different strategies that children had used. While children solved the problems at their desks, Ms. Sullivan interacted with individual students about their strategies. When interacting individually or during the class discussion with children, Ms. Sullivan consistently asked for complete and comprehensible explanations of why students did what they did. She frequently asked students to think about their strategies and to figure out whether the strategy might be simplified. For example, 1 student multiplied 9×9 by putting out 9 ten bars and then counting 9 of the units in each 10 bar. In other words he counted the individual units by one 1, 2, 3, . . . 78, 79, 80, 81. He said that he wanted to do something with 90 but it did not work. Ms. Sullivan probed what he was thinking. Her questioning provided sufficient scaffolding that the student recognized that the 9 ten bars was 90, and based on her questioning, he saw that he could count back 9 from 90 rather than counting from 1 to 81.

Ms. Carroll

Ms. Carroll shared Ms. Sullivan's beliefs about how children learn mathematics. In fact, she used the word *constructivist* to describe her philosophy of children's learning and characterized her perspective as "the opposite of the empty vessel approach." She described her role as a teacher in terms of listening to students and providing them with opportunity to invent their own solutions to problems. She did not characterize invention as entirely an individual activity. Rather she described a more social constructivist perspective in which students sharing their strategies for solving problems played a major role.

> We talked about listening to their thinking and when you do you value them. And when they come up with their own strategies and things, they're so clever. . . . Active involvement is really the hook that gets the kids going. It's not so much teaching as sharing knowledge. There's so much sharing that goes back and forth between the children and myself and then among the children too. It's very constructivist, and it's like taking what we have and moving on ourselves.

Ms. Carroll expressed a deep concern for helping children to develop understanding, which she characterized in terms of building connections: "I just think that's so important. It's not building blocks, but it's tinker toys. How you can connect here and there and everywhere and that's when you get a lot of those 'ah huhs.'"

Ms. Carroll's perspective of her own change. Ms. Carroll evidenced a perspective on her own learning that was quite different from Ms. Sullivan's, particularly in regard to knowledge of students' thinking. When asked whether she had changed since the professional development program ended 4 years ago, Ms. Carroll responded: I hope I'm growing, and I do think I've changed. But when asked specifically whether her understanding of children's thinking had changed, she reported that she was more aware of the importance of listening to students and that she listened more to them, but she never talked specifically about changes in her understanding of student thinking: "I realize how much more communication is. It's so important, I try to listen more. I try to let them go and sort of run with it."

Ms. Carroll thought that there was more to learn about children's thinking, but she felt that she would acquire that knowledge by taking classes and reading rather than through interactions with her own students. She tended to view knowledge of children's thinking as a fixed body of knowledge that was generated by other people. She did not appear to view her role as generating and testing hypotheses about children's thinking. She never talked about developing a deeper understanding of children's thinking by interacting with her own students.

> No, I don't think you ever get enough. I think it would be smart to take some more classes. I suppose if you read or you just reviewed your stuff that would help. Like the CGI handouts and things. Every so often I go back and look at them and say, oh yeah, I haven't tried, I haven't thought about that for a while. So that helps when I review that. Is that like reflection then? I suppose. Except I'm not thinking about it, I'm going to look it up. I've done

a lot of reading with all that brain research and constructivist theory and what else? Oh more of the seven intelligences.

During the professional development program, Ms. Carroll had been in a school in which most of the other primary grade teachers were participating in the program. Ms. Carroll described the community support for her learning. However, there were important differences in the way she described the teachers as a community of learners and the way she described the community of students in her class. She characterized the community of teachers as sharing her values and providing motivation to read and study, rather than as a community in which to engage in practical inquiry and a forum to share discoveries about students' thinking: "It's harder to make Yourself read and study and be challenged ourselves. But if you're in an organized group, like the support group, like the CGI classes really were, it's much easier to keep going and expand your knowledge."

Specificity of change. Ms. Carroll spoke passionately about the importance of listening to children and about how she valued their thinking, but she talked about children's thinking in general terms.

> It [CGI] made me take the time to look at what the kids were doing and listen to their words, and . . . it's not the product it's the process, it's getting to it and spending time.

> With CGI you see the growth through time . . . you value the child, and you don't value the test results or the finished product as much. You really look at the child, you have to value where the kids are and their thinking and moving them along.

Unlike Ms. Sullivan, Ms. Carroll seldom spoke about specific strategies that children used. The most specific descriptions she gave of student thinking focused on rather general strategies such as using mental math or not being able to count to 10. She never talked of the strategies children used in the detail that Ms. Sullivan did. The following quotes represent the most specific descriptions of student thinking in the interview:

> . . . figuring out where they are and what they know and who's comfortable with this kind of problem and who can work independently, and who can't count to 10 yet.

> And value where they are. Really value where they are and what there [their] thinking is. Because the child who's working on counting to 10. Well she was so happy when she got there.

> The kids who looked at those problems today and just used all sorts of mental math and answered the questions. It was fun to see the sparkle in their eyes. . . . I value listening to their thinking.

Classroom interaction. Ms. Carroll's classroom practices reflected her stated beliefs about how knowledge is acquired, both for her students and for herself. Students were provided opportunity to invent and share strategies for solving problems, but Ms. Carroll's interactions with students about the strategies they used did not provide her a basis for developing a deeper understanding of children's thinking. During the class observation, the children solved two multi-digit subtraction problems and five problems involving numbers less than 10. Some of the problems were from a book that Ms. Carroll read to the children, the others were written by the children. Ms. Carroll never demonstrated a strategy for any of the problems, and encouraged the children to solve them in their own ways. The children shared a variety of informal strategies. For example, 1 child subtracted 49 from 500 by first subtracting 50 and then adding back 1. Ms. Carroll always asked the students to explain how they solved a given problem. This was clearly a well-established class routine, and children explained their strategies as a matter of course. Unlike Ms. Sullivan, however, Ms. Carroll asked no question beyond "How did you get that?" We never observed her asking children to explain their strategy or clarify their explanations. One child's description of her strategy using the number line was so incomplete that it was impossible to know how she counted on the number line, i.e., whether she counted by tens or by ones. Ms. Carroll let the explanation stand and did not ask the child to elaborate. In the interview following the class, Ms. Carroll was unable to articulate how the child had solved the problem.

Summary of the Cases

Although the teachers in the two cases held similar beliefs about student learning, the differences in their conceptions about their own learning impacted their teaching practices in complex ways. The interactions in Ms. Sullivan's class afforded students greater opportunity to learn from other students, to learn to communicate their ideas more clearly, and to move more expeditiously to using more advanced strategies. The students in Ms. Carroll's class were not deprived of these opportunities. They had ample occasion to invent and discuss alternative solutions, and they were clearly learning, as demonstrated by their creative informal strategies for solving problems. However, Ms. Sullivan consistently asked her students to articulate their solutions and describe their rationale. She routinely interacted with students about the strategies they used and attempted to scaffold their use of more sophisticated versions of their strategies.

Perhaps the more profound differences were in the different opportunities that the two classes offered for the teachers to learn. We sensed a complex interaction between the beliefs and practices of the 2 teachers. Ms. Sullivan believed that she could learn from her students, and her classroom practices provided a context for her learning. Ms. Carroll did not perceive her classroom as a place for her own learning about student thinking, and her class interactions provided relatively little opportunity for such learning. It is not clear from these cases whether teachers construct classrooms in which they can learn from students because of their beliefs about engaging in practical inquiry to better understand student thinking, or whether their beliefs come from interacting with and learning from their students. We suspect that the truth lies somewhere in between: that teachers' practices and beliefs evolve interactively as they listen to and try to understand students' thinking.

Ms. Carroll had been listening to students for many years and we saw that she learned important lessons from listening to them. She learned that her students were capable of inventing strategies for solving problems, and she came to value their different solutions. Listening to students allowed her to sustain her practice over 7 years. Recognizing the value of her students' solutions provided her support and reinforcement for continuing to listen and provided students with opportunities to invent and discuss their own solutions. If her knowledge of student thinking had been entirely book learning, it is unlikely that it would have been sustained for such a long period. However, despite her years of listening, Ms. Carroll's knowledge of student thinking had not become generative. Simply listening to students talk about their thinking was not enough. When teachers struggle over extended periods to understand their students thinking, then their practices and beliefs begin to evolve together.

Support for Change

In our interviews with teachers we asked them to reflect on why they felt they were engaging with children's thinking in the ways that they reported. Because all of the teachers either reported that they had sustained their practice or had grown, we focused on why they thought that had occurred and what the supports and barriers were to their sustaining. Each teacher began the project in a school where the majority of teachers were participating in the professional development program. Many teachers reported that that level of support from colleagues was critical, in that it made the reform a school endeavor rather than a single teacher's endeavor. Many also reported that the long-term commitment of the research team made a difference to their continuing the implementation. However, the teachers differed in their collegiality and they made different choices in seeking out support.

To understand the different perspectives on collegial support, we focus on two schools, Malcolm and Oakwood Elementary Schools. In both schools, the majority of the kindergarten through third-grade teachers participated in the professional development project. Each school had a cadre of Level 4B, 3, and 2 teachers. Within both schools, some teachers developed high levels of collegial support while other teachers worked primarily alone. However, relationships and collegial support developed in different and sometimes surprising ways within the two schools.

At both schools, one grade level group of teachers engaged in daily, ongoing support of one another related to their teaching and learning of mathematics. Three fourth- and fifth-grade teachers[5] at Oakwood Elementary and 3 first-grade teachers at Malcolm Elementary planned together, questioned each other, shared articles with each other, talked about tasks, and talked about students, all focused on learning more about children's mathematical thinking in their classrooms. The teachers in these collaborative groups felt that continuing the reform without this level of support would be difficult. Although these teachers stated they would continue CGI without the support, each teacher commented that the support helped them learn. Ms. Sage, a Malcolm teacher commented,

> Particularly having [Ms. Abbot and Ms. Baker] here, I mean, we really talk a lot about what we are doing. . . . I think it is increasing my knowledge base. But I also think I am a better practitioner . . . we've had to build our own structure, but I think it was awfully important that you have something, I am just not sure where I would be without it. I would not be as far along as I am. I mean, I am not sure I would be able to go back to the workbook kind

---

[5]Each of these teachers was either a third- or first-grade teacher throughout the project. They changed grades together after the project ended.

of thing, even if I'd been alone, but I know I wouldn't be where I am without the support that I've had. (Ms. Sage, Level 4B)

The teachers within these collegial relationships constituted their community broadly. At the same time that they each talked about their own daily small group interactions, they also described learning from their students, working with university experts or district mentors, and reading articles. Ms. Castelbury, a Level 4B Oakwood teacher, talked explicitly about how working with her students and then talking with her colleagues helped her learn.

> We have to rely on each other more and that's okay. I don't think we've stopped doing that, in fact, I think we have become even more adamant that we need to be doing CGI kind of stuff. Because the reminders are always there within the kids. . . . It happens because when there's, because it's such a dynamic, it's such a dynamic thing that goes on when you solve a problem in the class. There are so many minds that are working on it and there's always something new to see happen and nothing ever is the same twice. So you just keep adding onto your bank of possibilities, things that kids can do. I mean, there are some trends, but I think they're the ones that kind of open up your, keep reminding you that there's still another way to think about this problem. And then through our discussion with other . . . you know, if I talk with [Ms. Madeline or Ms. Cohen] about what they see, then you even hear more.

All the teachers participating in these two collaborative groups were at Level 4B at follow-up and one of the teachers was the only teacher who developed from a Level 3 to a Level 4B.

Although the Level 4B teachers saw collegial support as helpful, the other teachers in these same schools did not develop the same level or type of collegial support. Furthermore, being a part of a supportive relationship did not mean the same thing to each teacher within the different groups. For example, the grade-level planning group at Malcolm involved a Level 2 teacher. There was a noticeable difference in how the Level 2 teacher, Ms. Conti, viewed the grade-level collaboration and the way the other teachers involved viewed it. Ms. Conti participated in the weekly planning meetings and considered her colleagues supportive. Even so, she rarely engaged with the teachers about children's thinking during their planning meetings. She would take the tasks that the group developed and use them in her classroom in a way that made sense to her, but she never really engaged with the community beyond sharing tasks. Yet Ms. Conti described the collaboration as helpful, especially in terms of sharing ideas of what to do in the classroom.

> We have a team concept here, so we're planning our units together, and when we plan our units we plan with the language and math and science and social studies, we plan with all these things in mind. So I think for us, it's not as bad you know, because we still have support. In essence, we still have each other, helping each other and coming up with some new ideas and things like that. . . . Because, I mean, after all we do have [Ms. Abbot and Ms. Baker]. Because we still share, we still work out ideas together, so it's not like we were left hanging here. If I was teaching all by myself, without any of these other people around, and I still wanted to do it, yeah, I'd probably be looking for support.

Ms. Conti remained at a Level 2, even after more than 6 years of working with a group of 3 other Level 4B teachers.

Opportunities existed for other teachers in each of these schools to collaborate with colleagues within and across their grade level, but they made other decisions regarding collaboration and support. Some teachers felt they had the support they needed from their colleagues, but this support focused on social support rather than sharing children's thinking. For example, when asked about whether she talked with any of her colleagues about CGI, Ms. Carroll, Level 3, stated, "not much." She told us that she thought it was most important that teachers received the support in the first couple of years after the initial inservice, "like we did," but that she could use help thinking of activities.

> I was talking about level of support. It's a huge change. We did so much planning together. And we'd say, oh, let's do geometry, and then we'd get together and say, oh, this is a good book, this is a neat thing. . . . Sharing the camaraderie, and not only that, but man, I can't think of this stuff all by myself but I don't really know that much of what other people are doing. I don't get into their rooms very much. I know that I need to get myself in contact with the people from here and spend more time with them. I have to get myself going. But this year we've been busy.

Many of the Level 2 and 3 teachers made similar decisions. They reported talking with their colleagues on occasion, sometimes even sharing what one of their students was doing. But they neither talked with each other regularly nor felt like they were missing out on anything. Overall, these teachers felt they had support.

Other teachers found little collegial support within their schools and looked outside. This was especially true for the Level 4B teachers. For example, Ms. Sullivan told us that she needed colleagues, but not just any colleagues. She could go next door and talk to the teacher about her students' thinking, but she saw this more as helping than support. "I'm really not sure it's the bouncing of what kids are doing with another colleague as much as I think it helps to bounce it off someone who really has knowledge about kids' thinking." Ms. Sullivan looked for support from teachers with knowledge of children's thinking that could push her. Ms. Sullivan's learning community came from outside of the school (as she took on workshops for other teachers) and from inside of her classroom.

Summary

It was not surprising to find that some Level 4B teachers created mutually interactive relationships with their colleagues to support their practical inquiry. More interesting was the possibility that teachers might become generative by engaging in one of these communities. The only teacher who became a Level 4B teacher after the project ended did so with support from her grade-level colleagues and a research group focused on children's mathematical thinking. In both of these learning communities, she tested out her knowledge of children's thinking, discussed it, revised it, and tried it again.

Also intriguing was the fact that while all of the teachers engaged in generative change were committed to creating their own learning communities, these communities differed with the specific needs of the teacher and in relation to the other communities within which the teachers worked. Some created elaborate learning communities in which they engaged in inquiry, constantly sharing students' problem-solving strategies and discussing what those strategies suggested about the students' understanding. Other teachers engaged in inquiry about the development of children's thinking as they led professional development opportunities for their colleagues, or for teachers in schools other than their own. In contrast, the teachers not engaged in generative growth did not use their relationships with their colleagues as opportunities for engaging in inquiry about children's thinking. More typically, they used collegial relationships to provide moral support.

Discussion

Four years after participating in a professional development program focused on the development of children's mathematical thinking, all 22 teachers maintained some use of children's thinking and 10 teachers continued learning in noticeable ways. Our examination of how these teachers engaged with children's thinking and how they found support for their work revealed insights about generative growth and sustainability of changed practice. These new understandings also have important implications for professional development.

Generativity

Only the teachers who listened carefully to the details of their children's mathematical thinking and then used what they learned to make ongoing instructional decisions (Level 4B) demonstrated engagement in ongoing learning about the development of children's thinking. As we saw in the case of Ms. Sullivan, such teachers not only understood children's thinking in organized, principled ways and talked specifically about their children's thinking, they also viewed knowledge of children's mathematical thinking as their own to add to and adapt. Level 4A teachers had similar knowledge about the structure and detail of children's mathematical thinking, but they did not see the knowledge as something to continue to learn about. These teachers successfully used children's thinking in their classrooms and they sustained their practice. However, because they did not see the knowledge as their own, their classrooms did not become places for their own learning. From these observations we speculate that organized knowledge about children's mathematical thinking is not enough to engage teachers in generative change. Rather, for change to be generative, teachers need to develop a view that the knowledge of children's mathematical thinking is their own to create, adapt, and investigate.

Sustaining Practice

Listening to children's mathematical thinking enabled teachers to sustain their practices. From listening to children, Level 3 teachers learned that their students were capable of using a variety of strategies, and they came to value their students' solutions. This valuing supported their continued listening. Once teachers valued their students' thinking, they saw benefit in asking their children how they solved problems. Consequently, they continued to provide opportunities for students to use a variety of strategies and to talk about their thinking. However, listening to children did not ensure that the teachers would learn more about children's thinking.

The Level 3 teachers sustained their change but had not become generative. They could articulate the problem types and the strategies children used, but they did not discuss or make use of connections among problems and strategies in ways that highlighted the principled ideas underlying them. The Level 3 teachers provided their children with opportunities to solve problems in their own ways and opportunities to build on their own thinking. However, they did not push their children to be more detailed in their descriptions, to compare strategies, or to use other processes that would enable students to build more readily on their current thinking.

Learning with Understanding

When we began this paper we outlined three elements of learning with understanding: knowledge that is generative, connected, and seen as one's own. Drawing on these elements we are coming to understand generative change in teachers as the link between connectivity and making knowledge one's own. Generativity seems to relate to the relationship between the structure of teachers' knowledge (the ways in which they see their ideas as related) and their view of that knowledge as their own to create, adapt, and change. We observe interplay between what teachers do in the classroom, the information that they gather there, and the ways in which they make sense of that information. However, rather than being random or simply reflective, that interplay is focused and deliberate. It extends their substantive understanding of their own students mathematical thinking and insights into a broader picture of the development of mathematical thinking.

Teachers who have developed a framework for analyzing students' mathematical thinking use this structure as they assimilate what they hear from students. The framework helps teachers know what to listen for and how to connect what they hear to their other knowledge. They have a means to add detail and restructure their knowledge. If teachers view the knowledge as something they can add to and change, the framework provides the connections that allow knowledge to become generative. However, if teachers do not see the knowledge of children's thinking as their own, the frameworks as adaptable, or the knowledge as connected, their ability to develop new understanding is limited.

Generative Change in the Context of Collaboration

Learning with understanding and generativity develop within the context of communities. All of the teachers involved in this project were passionate about the role their colleagues played in their learning. Our data focused primarily on how teachers sustained their relationships with their colleagues after the professional development project had come to an end. We talked to teachers about how their collegial relationships evolved and the ways in which the relationships met their perceived needs. Like other researchers, we found that helping schools become places where teachers learn is a difficult task.

Hargreaves (1992, 1994) talks about both the forms and the content of teacher cultures. He presses readers to resist thinking about the forms and content of collaborative relationship as a dichotomy (either they exist or they do not) but rather to think about them on a continuum. Our data provide additional support for Hargreaves' conceptualization of collaboration. Yet our data also causes us to think more critically, as does Hargreaves himself, about collaboration. Our data suggest that the forms and content of collaboration are interrelated. The teachers at Level 4B who engaged in generative change shaped both the form and content of their collaborative relationships to permit practical inquiry about children's mathematical thinking. These teachers who sought out opportunities to learn more about the development of children's mathematical thinking adapted existing forms of collaboration, either inside or outside of their schools, to meet the needs of the new content focus. The teachers involved in sustaining their practice but not in generative growth did not see the need for participating in new forms of collaboration focused on practical inquiry about children's mathematical thinking. These teachers often drew on their individual relationships for social and occasional substantive support rather than building group collaboration.

Professional Development

Although we have more to learn, the 22 teachers engaged in this project provided us with an opportunity to learn about generative change and begin to understand why some teachers may become engaged in ongoing learning while others struggle. The teachers also provided insights about professional development. This study provides support for using student thinking as the basis of professional development, while also raising broader issues about the design of professional development.

Focusing on student thinking proved to be a valuable mechanism for engaging teachers in generative change. Our results support and extend Knapp and Peterson's (1995) findings that teachers who focus on the principled ideas underlying children's mathematical thinking sustain themselves. Generative growth occurred for teachers who perceived themselves as learners, creating their own understandings about the development of student thinking. We propose that it is the engagement with student thinking that allowed teachers to develop understanding and connect ideas. As teachers engage with student thinking, they think about their daily work, about substance, content, and process, and about their own students. They come to see that they can learn through working with their own students in their own classrooms; they receive continual feedback as children discuss their thinking. Teachers can create learning communities that involve their students and their colleagues: they can learn as they engage with their students and continue that learning as they engage with their colleagues. In listening to their students and then talking about it with their colleagues, teachers are not simply sharing; they are building principled knowledge on which to base their ongoing instructional decisions.

Professional development focused on student thinking is not unique to CGI. A number of projects currently engage teachers in learning about the development of children's thinking. Many of these projects explicitly engage teachers in practical inquiry in their classrooms and their school communities (for example see, Schifter, 1998; Lehrer & Schauble, 1998; Warren & Rosebery, 1995). In combination, this work and our current findings provide considerable evidence to support a focus on student thinking as a means for engaging teachers in generative growth.

As we begin to accumulate knowledge about student thinking as the focus for teacher learning, we are finding that engaging in generative growth is not a solitary endeavor. Our data suggest that teachers' collaborations can support their learning. Engaging teachers with student thinking requires thinking about how teachers can develop collaborations. This may not seem like a new idea. However, we are suggesting that the content and form of the professional development must be consistent. Professional development that encourages teachers to listen to their students and to learn from them requires opportunities for the inquiry to be fostered. Teachers need time to develop relationships with others that they can talk with in ways that meet their needs and push their thinking. However, our data suggest that institutionalizing collaborative work or mandating practical inquiry will not work. This point became clear as we considered the experience of a Level 2 teacher involved in a significant amount of group collaboration. While the 3 other teachers in the collaboration engaged in practical inquiry regularly, this teacher did not. Rather, she participated in a way that met her needs and fit with how she thought about the teaching and learning of mathematics. We speculate that once teachers reach a Level 3, participating with other teachers engaged in practical inquiry will provide them an opportunity to develop further. As they observe other teachers adapt, create, and challenge their own thinking they could see that they could learn from their own students in their classrooms and also from their colleagues. They could also become aware that knowledge was their own to continually recreate.

More of an effort needs to be made to engage teachers in professional development that supports their ongoing learning and simultaneously provides opportunities for teachers to create collaborations with their colleagues. Doing one or the other alone was not enough to enable most teachers to become generative once the professional development ended. Although developing schools as learning communities remains an area for further work, our work can guide inquiry about the relationship between form and content of collaboration and the teachers' role in constituting the collaboration when the intention is to engage teachers and school communities in ongoing learning.

Conclusion

We are not proposing that analyzing children's thinking is the only avenue for teachers' growth to become generative; however, this focus does have characteristics that support generative growth. Children's thinking is available to teachers in their own classrooms daily: there are regularities in the strategies that children describe and principled ideas about these regularities. In addition, this focus allows teachers ways to talk with each other about their classrooms and their students. Teachers can create communities of learning that focus on children's thinking. Teachers who are focused on their children's thinking create learning communities that include their classrooms. Consequently, these broadly inclusive learning communities encourage teachers to engage in inquiry focused on children's mathematical thinking, and they share this inquiry with their students as well as with their colleagues.

Notes

An earlier version of this article was presented at the annual meeting of the American Educational Research Association, San Diego, April 1998. The research reported in this paper was supported in part by a grant from the Department of Education Office of Educational Research and improvement to the National Center for Improving Student Learning and Achievement in Mathematics and Science (R305A60007–98). The opinions expressed in this paper do not necessarily reflect the position, policy, or endorsement of the Department of Education, OERI or the National Center.

We thank the participating teachers and gratefully acknowledge the contributions of the SSGC research group at UCLA.

References

Ashton, P. T., & Webb, R. B. (1986). *Making a difference: Teachers sense of efficacy and student achievement.* White Plains, NY: Longman.

Barnett, C. S., & Sather, S. (1992). *Using case discussions to promote changes in beliefs among mathematics teachers.* Paper presented at the annual meeting of the American Educational Research Association, San Francisco.

Borko, H., & Livingston, C. (1989). Cognition and improvisation: Differences in mathematics instruction by expert and novice teachers. *American Educational Research Journal, 26,* 473–498.

Borko, H., Mayfield, V., Marion, S., Flexer, R., & Cumbo, K. (1997). Teachers' developing ideas and practices about mathematics performance assessment: Successes, stumbling blocks, and implications for professional development. *Teaching and Teacher Education, 13*(3), 259–278.

Borko, H., & Putnam, R. (1996). Learning to teach. In D. Berliner & R. Calfee (Eds.), *Handbook of educational psychology* (pp. 673–708). New York, NY: MacMillian.

Brown, A., & Campione, J. (1996). Psychological theory and the design of innovative learning environments: On procedures, principles, and systems. In L. Schauble & R. Glaser (Eds.), *Innovations in learning* (pp. 289–326). Hillsdale, NJ: Erlbaum.

Carpenter, T. P., Fennema, E., & Franke, M. L. (1996). Cognitively Guided Instruction: A knowledge base for reform in primary mathematics instruction. *Elementary School Journal, 97*(1), 1–20.

Carpenter, T. P., Fennema, E., Peterson, P. L., & Carey, D. A. (1988). Teachers' pedagogical content knowledge of students' problem solving in elementary arithmetic. *Journal for Research in Mathematics Education, 19,* 385–401.

Carpenter, T. P., Fennema, E., Peterson, P. L., Chiang, C. P., & Loef, M. (1989). Using knowledge of children's mathematics thinking in classroom teaching: An experimental study. *American Educational Research Journal, 26*(4), 499–531.

Carpenter T. P., & Lehrer, R. (1999) Teaching and learning mathematics with understanding. In E. Fennema & T. R. Romberg (Eds.), *Mathematics classrooms that promote understanding.* Mahwah, NJ: Erlbaum.

Cobb, P., Wood, T., & Yackel, E. (1990). Classrooms as learning environments for teachers and researchers. *Journal for Research in Mathematics Education Monograph, 4,* 125–146.

Dana, T. M., Campbell, L. M., & Lunetta, V. N. (1997). Theoretical bases for reform of science teacher education. *Elementary School Journal, 97*(4), 419–432.

Fennema, E., Carpenter, T. P., Franke, M. L., Levi, L., Jacobs. V., & Empson, S. (1996). A longitudinal study of learning to use children's thinking in mathematics instruction. *Journal for Research in Mathematics Education, 27,* 403–434.

Franke, M. L., Fennema, E., & Carpenter, T. P., (1997). Changing teachers: Interactions between beliefs and classroom practice. In E. Fennema & B. Nelson (Eds.), *Mathematics teachers in transition.* Mahwah, NJ: Erlbaum.

Franke, M. L., Fennema, E., Carpenter, T., Ansell, E., & Behrend, J. (1998). Understanding teachers' self-sustaining change in the context of professional development, *Teaching and Teaching Education, 14*(1), 67–80.

Fullan, M. (1991). *The new meaning of educational change* (2nd ed.). London: Cassell.

Giroux, H. A. (1988). *Schooling and the struggle for public life.* Minneapolis: University of Minnesota Press.

Goldenberg, C., & Gallimore, R. (1991). Local knowledge, research knowledge, and educational change: A case study of early Spanish reading improvement. *Educational Researcher, 20*(8), 2–14.

Greeno, J. (1988). *Situations, mental models and generative knowledge.* (Rep. No. IRL 88-0005). Palo Alto, CA: Institute for Research on Learning.

Hargreaves, A. (1992). Cultures of teaching: A focus for change. In A. Hargreaves & M. Fullan (Eds.), *Understanding teacher development.* London: Cassell and New York: Teachers College Press.

Hargreaves, A. (1994). *Changing teachers, changing times: Teachers' work and culture in the postmodern age.* New York: Teachers College Press.

Hiebert, J., & Carpenter, T. P. (1992). Learning and teaching with understanding. In D. Grouws (Ed.), *Handbook of research on mathematics teaching and learning* (pp. 65–97). New York: MacMillian.

Knapp, N., & Peterson, P., (1995). Teachers' interpretations of CGI after four years: Meanings and practices. *Journal for Research in Mathematics Education, 26,* (1), 40–65.

Lehrer, R., & Schauble, L. (1998). *Modeling in mathematics and science.* Unpublished manuscript. Wisconsin Center for Educational Research: Madison, WI.

Lieberman, A., & Miller, L. (1990). Teacher development in professional practice schools. *Teachers College Record, 92*(1), 105–122.

Leinhardt, G., & Greeno, J. C. (1986). The cognitive skill of teaching. *Journal of Educational Psychology, 2,* 75–95.

Little, J. W. (1993). Teachers' professional development in a climate of educational reform. *Educational Evaluation and Policy Analysis, 15,* 129–151.

McLaughlin, M. W., & Oberman, I. (1996). *Teacher learning: New policies, new practices.* New York: Teachers College Press.

McLaughlin, M. W., & Talbert, J. E. (1993). *Contexts that matter for teaching and learning: Strategic opportunities for meeting the nation's educational goals.* Stanford, CA: Center for Research on the Context of Secondary School Teaching, Stanford University.

Peterson, P. L., McCarthey, S. J. & Elmore, R. E. (1996). Learning from school restructuring. *American Educational Research Journal, 33*(1). 119–153.

Richardson, V. (1990). Significant and worthwhile change in teaching practice. *Educational Researcher, 19*(7), 10–18.

Richardson, V. (1994a). *Teacher change and the staff development process: A case in reading instruction.* New York: Teachers College Press.

Richardson, V. (1994b). Conducting research on practice. *Educational Researcher, 23*(5), 5–10.

Richardson, V., Anders, D., Tidwell, D., & Lloyd, C. (1991). The relationship between teachers' beliefs and practices in reading comprehension instruction. *American Educational Research Journal, 28*(3), 559–586.

Sarason, S. (1990). *The predictable failure of educational reform.* San Francisco: Jossey-Bass.

Sarason, S. B. (1996). *Revisiting "The culture of the school and the problem of change."* New York: Teachers College Press.

Schifter, D. (1997 April). *Developing operation sense as a foundation for Algebra.* Paper presented at the Annual Meeting of the American Educational Research Association, Chicago.

Schifter, D., & Fosnot, C. T. (1993). *Reconstructing mathematics education: Stories of teachers meeting the challenge of reform.* New York: Teachers College Press.

Schifter, D., & Simon, M. (1992). Assessing teachers' development of a constructivist view of mathematics learning. *Teaching and Teacher Education, 8*(2), 187–197.

Secada, W. G., & Adajian, L. B. (1997). Mathematics teachers' change in the context of their professional communities. In E. Fennema & B. S. Nelson (Eds.), *Mathematics teachers in transition* (pp. 193–219). Mahwah, NJ: Erlbaum.

Sherin, M. G. (1997, April). *When teaching becomes learning.* Paper presented at the Annual Meeting of the American Educational Research Association, Chicago.

Shulman, L. S. (1986). Those who understand teach: Knowledge growth in teaching. *Educational Researcher, 57*(1), 1–22.

Simon, M., & Schifter, D.). (1991). Towards a constructivist perspective: An intervention study of mathematics teacher development. *Educational Studies in Mathematics, 22,* 309–331.

Tharp, R. G., & Gallimore, R. (1988). *Rousing minds to life.* New York: Cambridge University Press.

Warren, B., & Rosebery, A. S. (1995). Equity in the future tense: Redefining relationships among teachers, students and science in language minority classrooms. In W. Secada, E. Fennema, & L. Adajian (Eds.), *New directions for equity in mathematics education* (pp. 298–328). New York: Cambridge University Press.

Wood, T., Cobb, P., & Yackel, E. (1991). Change in teaching mathematics: A case study. *American Educational Research Journal, 28*(3), 587–616.

SESSION 6

Observing How Knowledge is Constructed in Mathematics Classrooms

The NCTM *Professional Teaching Standards* calls unprecedented attention to the "discourse" of mathematics classrooms, as embodied in three standards: Teacher's Role in Discourse, Students' Role in Discourse, and Tools for Enhancing Discourse. An unfamiliar term to many, discourse is used to highlight the ways in which knowledge is constructed and exchanged in classrooms. Who talks? About what? In what ways? What do people write down and why? What questions are important? Whose ideas and ways of knowing are accepted and whose are not? What makes an answer right or an idea true? What kinds of evidence are encouraged or accepted?*

In classrooms where students are exchanging mathematical ideas, teachers are faced with the challenge of determining what to do with students' ideas once they have been aired. These challenges are significant for all teachers, and through observations and discussion, you can help teachers learn how to attend to and make pedagogical decisions based on students' mathematical ideas.

In this session you shift your attention from the mathematics content of a lesson to the sense students are making of the content and the way the teacher works with students' mathematical ideas. Using the Learning & Pedagogy Observation Guide, you will focus your classroom observations on the teacher's pedagogical moves *in relation to the students' mathematical ideas* rather than focusing on the students or the teacher alone. This interplay between students' ideas and teacher's pedagogical moves is at the heart of classrooms in which generative learning is taking place.

You will continue to explore ways in which the content and modes of communication of your post-observation conferences with teachers might contribute to teachers' ongoing generative learning. You will also consider the role of inquiring collaboratively with teachers about the mathematical thinking that takes place in their classrooms. Such curiosity about students' thinking can fuel productive discussions about how to probe students' mathematical ideas and, through those discussions, support teachers' generative learning.

---

*excerpt from Ball, D. (1991). What's All This Talk About "Discourse"? *Arithmetic Teacher*, 44–48. Copyright ©1991 by the Association for Supervision and Curriculum Development. Reprint with permission from the publisher. All rights reserved.

"Canceling" Zeroes

A. In the next videotaped clip, you will see a fifth grade classroom exploring the question of "canceling" zeroes when looking for an equivalent fraction. For example, is it O.K. to "cancel" the zeroes in $\frac{10}{30}$ to get the equivalent fraction, $\frac{1}{3}$? To help you make sense of the mathematics you will be seeing in this classroom episode, please do the following mathematics assignment.

Below is a set of fractions, all of which have zeroes in the numerators and denominators. Sort these fractions into at least 3 categories. In one category, put the fractions where you do not get equivalent fractions when you cross out zeroes. In the other categories, put fractions where you <u>do</u> get equivalent fractions once you've crossed out the zeroes. For these other categories, describe what mathematical principle allows you to cross out the zeroes. Think of some rules to determine when "canceling" zereos works.

$$\frac{10}{30} \qquad \frac{2001}{6003}$$

$$\frac{10}{101} \qquad \frac{204}{408}$$

$$\frac{101}{201} \qquad \frac{202}{205}$$

$$\frac{2010}{5010} \qquad \frac{6200}{8700}$$

B. Analyzing Discourse: In the next session, when you view the videotape of a fifth grade classroom, you will be focusing on the mathematical discourse that takes place among students and between the teacher and his students. To prepare you to attend to these things, please reread *What's All This Talk About "Discourse"?* by Deborah Ball (Reading 3) and find some examples of how Ball works with students' ideas. Be ready to share them during the next session.

READING 8

Freedom to Explore[1]
Friday: "Is it O.K. to Cancel those Zeroes?"

Teacher: Mr. Punzak
Total Time: 23 minutes
Tape Version: Final © 1997

The video begins with Mr. Punzak at the board going over a homework problem.

~00:00

1	Mr. Punzak:	Divide 400 . . .
2	Class:	400
3	Mr. Punzak:	. . . by 30.
4	Student:	Can I do it?
5	Mr. Punzak:	And what's it say to do, Nicky, with the remainder?
6	Nicky:	. . . in fraction form.
7	Mr. Punzak:	Put the remainder in fraction form. OK?
8	Girl:	Can I go up to the board, now?
9	Mr. Punzak:	Sure.
10	[*A girl comes to the board.*]	
11	Mr. Punzak:	Seminta, could you turn around and make sure this is . . . you agree with this? [*The girl calculates a remainder of "10" for the long division problem, while talking to herself.*] All night, now what do I do next, Seminta?
12	Seminta:	You uhm, put the, ah, remainder in fraction form.

[*Seminta at the chalkboard writes 13 and $\frac{1}{3}$.*]

---

[1]From Mathematical Inquiry Through Video BBNT Solutions LLC copyright 1997

14	Mr. Punzak:	OK, 13 and one-third. Gee, I don't know where you got that. I can guess, but I'd rather you told me.
15	Seminta:	Well . . . OK.
16	Mr. Punzak:	I see the 13. But how'd you get the . . . [*circles "$\frac{1}{3}$"*] one-third?
17	Seminta:	OK, [*talks to herself as she writes "10" on the board, checks back at her paper, and finishes writing "$\frac{10}{30} = \frac{1}{3}$."*]
18	Seminta:	OK, well, what I did here is, I reduced it.

~01:00

19		[*Seminta turns from the board and faces Mr. Punzak.*]
20	Mr. Punzak:	[*to Seminta*] Ten-thirtieths, you reduced to one-third?
21	Seminta:	Yea.
22	Mr. Punzak:	[*to the class*] Does that make, make sense to you, everybody?
23	Students:	Yea.
24	Mr. Punzak:	OK. Good. So, if you didn't get thirteen and one-third, or thirteen and one-third is not the answer on the paper you are looking at, put a circle around them please. Did anybody circle one . . . no? Great [*starts erasing the board*].

~01:45

25		[*The view of the class shows students are at their desks, arranged in three groups. One cluster is near the chalkboard and consists of female students. The other two, toward the back of the room, are made up of male students.*]
26	Boy:	Mr. Punzak, wait. What if they, what if you got thirteen . . . instead of thirteen and one third, what if you got thirteen and ten-thirtieths. Is that the same as . . . [*The student's voice becomes inaudible as a girl speaks.*]
27	Girl:	Yea.
28	Girl:	It's the, it's the same thing. Just take out the zeroes.
29	Mr. Punzak:	OK? [*speaks while writing on the board*] Is this the same as . . .

$$\frac{10}{30} \stackrel{?}{=} \frac{1}{3}$$

Figure 1. Equivalent fractions problem with final zeroes in numerator and denominator.

30	Mr. Punzak:	This is the question you're asking? Is this the same? These two are the same? [*turns to the class*]
31	Tira:	When you divide something into thirds, that, you're putting it in three equal parts [*says last three words with equal stress and a gesture marking each*], and, in thirty, ten is one of the three equal parts. Ten plus ten plus ten equals thirty. And one plus one plus one equals . . . three [*smiles*].
32	Kate:	[*very faintly, before she's on camera*] May I say something else?
33	Mr. Punzak:	Kate?
34	Kate:	[*rises and approaches the chalkboard*] You got to cancel the zeroes and then it's still one-third. Or divide them both by ten.
35	Mr. Punzak:	OK
36		[*inaudible agreements from a few people in Kate's group of desks as Kate returns to her seat*]
37	Kate:	And that wouldn't change the lines.

~02:30 Mr. Punzak's first counter example

38	Mr. Punzak:	All right . . . Let me just ask you one thing, since you've said about canceling by zeroes, what about this. . . ? [*writes the following counterexample on the board*]

$$\frac{1\cancel{0}}{1\cancel{0}1}$$

Figure 2. Mr. Punzak's first counterexample: zeroes in different columns.

39	Mr. Punzak:	Could you just cancel the zeroes and call that one-eleventh? [*turns to the class*]
40	Students:	[*Various muffled no's, then "oh's." "wait, could you" and "ok now." A few students raise their hands.*]
41	Mr. Punzak:	Talk it over in your groups.
		Mr. Punzak reconvenes the class.
42	Mr. Punzak:	OK, is every group satisfied with their answers?
43	Class:	Yea.
44	Mr. Punzak:	All right, this group here, uhhh, Nora.
45	Nora:	When you have . . . ten . . . like, if you have ten-thirtieths, zeroes are both in the units place, but in this case, that one's in the units place [*meaning one zero*] and one's in the tens place.

46	Mr. Punzak:	Did you want to disagree with that Kate or . . .
47	Kate:	Well, I just have a slightly different thing.
48	Mr. Punzak:	OK.
49.	Kate:	You can't switch around the numbers because those, because if you put this here [*pointing to the 1 in 101*] and the zero there [*points to 101 indicating a switch between the "1" in the units place and the "0"*], then a hundred and ten isn't the same as a hundred and one.
50	Mr. Punzak:	OK. I can understand that. O.K Lisan did you want to add something to your group?
51	Lisan:	The numerator always has to go into the denominator for it, for you to make it, like . . .
52	Mr. Punzak:	For you to reduce it?
53	Lisan:	Yea . . . [*watches Mr. Punzak who is presumably writing*] I think. . .I'm not sure, but . . .
54	Mr. Punzak:	OK. I'm not sure what that means. [*Mr. Punzak repeats Lisan's words and writes his statement on the board.*] First before we get back to this Lisan, I want to check with that back group. Andy?
55	Andy:	OK. Our group said that the two zeroes always have to be in the same column. Like the ones column and the ones column. One zero can't be in the, it's, otherwise it'd be different. Cause one zero can't be in the ones column and the other zero be in, like, the thousand, thousandths column. It'd be all mixed up then [*waves his arms in the air to signify 'all mixed up'*].

~04:30 Mr. Punzak's second counter example

56	Mr. Punzak:	What if I had, so, you said something about them having to be in the ones column. What if I had one hundred and one over two hundred and one. Could I do that? They're both in the same column.

$$\frac{1\emptyset 1}{2\emptyset 1}$$

Figure 3. Mr. Punzak's second counterexample: zeroes in the middle column.

57	Students:	Yea.
58	Andy:	Yea.
59	Mr. Punzak:	I could do that, huh? [*Mr. Punzak writes on board, but can't see what he's written.*]

60	Andy:	No. Because eleven doesn't go into twenty-one.
61	Mr. Punzak:	Hmm?
62	Student:	Of course it does.
63	Student:	No it, no it . . .
64	Andy:	No it doesn't.
65		[*unintelligible voices*]
66	Boy:	Eleven can go into twenty-one just like . . .
67	Mr. Punzak:	Well, it looks like this is a new question., so let's have your little groups decide on this. Can you do that? [*pointing to the board*] Nicky? [*Mr. Punzak moves to Nicky's table.*]

Mr. Punzak joins the discussion at Nicky's table.

68	Nicky:	They, they have to be in, like, they have to be in the same column, but they also have to be in the far right.
69	Mr. Punzak:	In the far right column? The units column?
70	Nicky:	Yea. They., always have to be in the units column.
71	Mr. Punzak:	I wonder why that's true.
72	Nicky:	It just is.
73	Mr. Punzak:	It just is?
74		[*In the background, the girls are clustered at the board talking.*]
75	Joe:	What happens if you have two zeroes? Can you cancel them both out? If they're in the same columns—can you go like this? [*makes an X in the air with his pencil*] Make an "X?" I mean, would that work?
76	Mr. Punzak:	If they were, if they're two zeroes next to each other?
77	Joe:	. . . If it's like this. [*He writes on his paper as Mr. Punzak looks over.*] If it's like . . .
78	Joe:	It's an interesting way of looking at things.
79		[*We see Andy's hands tracing all the possible permutations for canceling the zeroes in $\frac{200}{700}$. As he speaks, he traces lines indicating that crossing out zeroes either in the same or different columns produces the same result.*]

Figure 4. Andy crosses out zeroes diagonally from the zero in the ones column of "200" to the zero in the tens column of "700."

80	Andy:	Like this . . . [*across both zeroes in 200*], like this . . . [*across zeroes in tens column*], like this . . . [*diagonal from top tens column zero to bottom ones column zero*], like this . . . [*across zeroes in ones column*], this . . . [*diagonal from top ones to bottom tens column zero*], this . . . [*diagonal from bottom ones to top tens column zero*].
81	Mr. Punzak:	[*unintelligible*] get an answer to that?

~05:30 The girls discuss whether canceling zeroes in Mr. Punzak's second counter example works

82		[*Focus shifts to Mr. Punzak at the blackboard where he is surrounded by female students. There is a lot of noise in the room.*]
83	Mr. Punzak:	The question is . . . is this right? [*Mr. Punzak points to the $\frac{101}{201}$ problem on the chalkboard with the zeros crossed out.*]
84	Sarah:	Yes it is.
85	Tira:	Yes it is. Because look. [*Several students try to get to the problem on the board.*]
86	Kate:	But you can't, but you can't
87	Tira:	No, look.
88	Kate:	Oh, come on.
89	Tira:	Because look, you said it would be half if this was two hundred and two.
90	Kate:	Yea.
91	Tira:	And this would be half if this was twenty-two. And it, it is right.
92	Girl:	It's the same thing.
93	Kate:	[*Motioning to the fractions eleven over twenty-one and one hundred and one over two hundred and one*] Eleven, twenty-onths is the same as a hundred and one, two hundred and onths? [*looking at Tira*] But why?

~06:00 Tira explains her position,

94		[*Mr. Punzak is now at the board. All of the students are seated except Tira.*]
95	Mr. Punzak:	OK. The question is, I'm going to write it again, is, one hundred and one, two hundred and onths [*As Mr. Punzak writes on the chalkboard, Tira approaches him.*], can you cancel those middle zeroes?
96	Students:	Oh yes.
97	Tira:	And. . . [*looks disapprovingly at whoever was calling out*]

98	Students:	Oh yes, you can! [*same student as before*]
99	Mr. Punzak:	Shh, shh, shh . . .
100	Tira:	And we said, uhm, yes, because, eleven plus eleven . . . [*writes the addition problem on the board*]
101	Girl:	[*finishing Tira's sentence*] Is twenty-two.
102	Tira:	. . . is twenty-two.
103	Girl:	[*same as before*] Yes, that's why [*unintelligible*]
104	Tira:	So it's, it's one twenty-oneth away from being half of 21. One hundred and one is one two hundred and oneth away from being half of two hundred and one. Do we get it? [*Tira turns from the board smiling.*]
105	Mr. Punzak:	Are you correct? Is that what you are asking me? You're asking me [*laughs*] to tell you whether you're correct or not? [*extremely softly*] I'm not going to tell you that.
106	Kate:	[*off camera*] Wait a sec, wait a sec.
107	Mr. Punzak:	[*louder*] Now we're going to listen to another group first. [*Kate has her hand up.*] Do you want to add on to this?
108	Kate:	[*unintelligible, possibly "yea but . . ."*]
109	Students:	No! No!

07:00 Students consider other canceling possibilities: canceling the ones.

110	Kate:	This isn't the answer but, we also . . . we found out that canceling the ones in the units place doesn't work [*points to the board*].
111	Nicky:	But I wanted to add on something. But if you're canceling out the zeroes you could cancel out the, uh, all the other numbers, too.
112	Lisan:	Well, uhm, I just think . . . that, uhm.
113	Nicky:	Oh, oh, oh! [*impatiently raises his hand, then lowers his hand before raising it again more patiently*] Sorry.
114	Lisan:	Uhm . . . that you could cross out. . . . Oh no, you couldn't cross out the ones because, uhm, one . . . one hundred and one is . . . isn't half of 'two hundred and one.
115	Mr. Punzak:	One hundred and one is NOT half of two hundred and one, so you CAN'T cancel out these things?
116	Lisan:	Yea.
117	Mr. Punzak:	If you cancel out these things you're left with one half.

118	Lisan:	No, you can cancel out the zeroes, but you can't cancel out the ones.
119	Mr. Punzak:	You can cancel out the zeroes. But you are willing to accept that, it equals eleven twenty-firsts?

~08:00

120	Lisan:	Yea. But, I don't think you cancel out the ones because a hundred and one two hundred and oneths aren't equal to one half.
121		[*back to Nicky, who had raised his hand while Lisan was speaking*]
122	Nicky:	Well like if, I'm just saying like if it, if it comes out to one half, I don't think, like, eleven twenty-firsts would be right. For some reason it's just like, 'cause . . .
123	Mr. Punzak:	So you're not, you're not satisfied with this but you're not sure why?
124	Nicky:	Yea. Kind of.
125	Mr. Punzak:	OK. OK. Tira.
126	Tim.	You said that, uhm, if it comes out one-half the answer would be eleven twenty-firsts. [*As she speaks, Tira moves across the room to the board.*] But how do you know that you cancel these out [*apparently referring to the final "01's" in $\frac{101}{201}$*] to get one half? And how do you know that you cancel THESE out [*pointing to middle zeroes*] to get, to get eleven twenty-firsts?
127	Mr. Punzak:	Well I'm confused here [*walks over to Tira's group*]. Didn't you over here say it was OK to do that?
128	Tira:	Yea, we said that. But he's saying that if it comes out to one-half, then the answer eleven twenty-firsts must not be right. But he doesn't know that you cancel out the on—, if you cancel out the ones or not. And, well WE don't think you do. That was our opinion. But he's, he's just assuming, like, he's just saying that you cancel those out, and so this must be wrong. He's not giving a reason why to cancel those out.
129	Mr. Punzak:	<u>OK. And tell me again your reason for canceling out these zeroes over here?</u>
130	Tira:	[*<u>moves to the board used earlier by her group</u>*] <u>Because zeroes are place holders. And so, if . . . since it's just the place holder, if you knock it out, it will still mean the same thing.</u>

~09:30

131	Mr. Punzak:	[*looks away from Tira to the whole class*] Is, is everybody in the room happy with this? Cancel out these middle zeroes? Willing to do that?
132		[*various answers from around the room*]
133	Kate:	I'm not sure.
134	Mr. Punzak:	Kate's not sure? OK.
135		[*Kate is now at the board.*]
136	Mr. Punzak:	[*to Kate*] Go ahead.
137	Kate:	When you have just the number a hundred and one, you take out the zero [*erases the zero*], you have eleven, and so it's completely changed. But with a fraction, this doesn't, eleven goes, goes into twenty-one the same as 101 goes into 201, so I think that it works to cancel the zeros, but I'm not sure.

Nora's insight about checking equivalence

138		[*Nora is raising her hand, anxious to say something.*]
139	Nora:	Oh, oh, oh, oh, oh, oh, oh, oh.
140	Mr. Punzak:	Nora?
141	Nora:	Wait, all we have to do [*looking at the board, twisting her body, and speaking softly*], this if it's OK to cancel zeroes . . . Is we have to do, a hundred and one divided by two hundred and one.
142	Student:	What?
143	Mr. Punzak:	She says, all we have to do to find out if it's OK to cancel the zeroes, is actually divide a hundred and one by two hundred and one.
144	Students:	Yea.

~10:30

145	Kate:	One hundred and one divided two hundred and one was decimal five-oh-two-seven-four-eight-seven-five. [*0.5027487*5] and eleven divided twenty-one was decimal five-two-three-eight-oh-nine [*0.523809*].
146	Mr. Punzak:	OK. It may be hard to read back there but these are two different decimals, for those two fractions, [*Kate speaks inaudibly at the board as she writes near the decimals she has just written.*] What does that tell you? Joe? [*inaudible response*] That they are not the same. They tell you that you can NOT cancel those middle zeroes.

Mr. Punzak's homework assignment

147	Mr. Punzak:	OK, we are going to leave this problem for the time being and solve it again, work on it again on Monday. But first, I'm going to give you a little clue. Remember back as to when we, when we first found out that you could cancel. We did something like ten over twenty. And we said that ten was equal to five times what?
148	Students:	Two
149	Mr. Punzak:	We prime factored it. Right?
150	Students:	Ohhh!
151	Mr. Punzak:	Why did we say that it was okay to cancel, here? That's your weekend clue.

SESSION 7

Building Mathematics Understanding: More Than the Sharing of Ideas

Given the importance of helping students develop their mathematical thinking, teachers need to structure their lessons so they can explore and probe their students' mathematical ideas. However, it is not enough for teachers to simply have students share their initial thoughts; they need to work with students' ideas in some depth so that they can understand the validity of the students' mathematical reasoning. When teachers have that depth of understanding, they can think most productively about what steps to take to pursue and build on students' ideas. When teachers continually probe and make sense of their students' thinking and use what they learn to plan their next pedagogical moves, they engage in generative learning.

In this session you will have a final opportunity of the course to work on the process of developing your eye for standards-based mathematics classrooms and your skills in using collaborative inquiry with teachers. These two strands weave their way through the session as you reflect on the classroom observations and post-observation conferences you did for homework and on the mathematical discourse that takes place in a math investigation in a 7th grade classroom. As you did in the last session, you will once again use the Learning & Pedagogy Observation Guide to focus your viewing.

Supporting Generative Learning

As administrators, you need to sift through a range of considerations—from factors that relate to the specific teacher's classroom practice to those that concern school and district priorities—as you make decisions about what to discuss with a teacher in a post-observation conference. For this homework assignment, you will focus on the actual mathematical understandings articulated by students, an important focus, given that the development of children's mathematical thinking is the goal of good classroom practice. When administrators inquire with teachers about students' mathematical thinking, they are drawing attention to its importance in a classroom and highlighting how critical a detailed understanding of students' mathematical thinking is to the pedagogical decisions teachers need to make. Such a focus on students' ideas lays the groundwork for teachers' growing capacity to engage in generative learning.

Your homework for the next session is to use the Learning & Pedagogy Observation Guide to observe one of the two teachers you have been following for this course. After your observation, conduct a post-observation conference in which you try out a collaborative inquiry approach among the ways you communicate with the teacher, with the idea of supporting the teacher's generative learning in mind.

1. To prepare for this observation, have a pre-observation conference with the teacher using the Pre-observation Conference Questions (p. 125).

2. As you conduct your observation and take notes with reference to the Learning & Pedagogy Observation Guide, keep track of questions about students' mathematical thinking that occur to you that you might like to ask the teacher during your post-observation conference.

3. Decide what questions and ideas related to student thinking might be fruitful for you and the teacher to investigate together.

4. Conduct a post-observation conference with the teacher and try out a collaborative inquiry approach where appropriate. Then consider the following:
   a. What ideas about the thinking of students in the class did you and the teacher find especially interesting to explore together? What made them interesting?
   b. What perspectives and additional "wonderings" about student thinking did you come away thinking about? Which do you think the *teacher* might continue thinking about on her or his own?

Pre-observation Conference Questions

In order to help you make sense of what you will be seeing when you do your classroom observations, plan to meet with the teacher prior to the observation and ask the following questions:

1. What topic will you and your students be working on in this lesson?

2. What do you plan to do in this lesson? (e.g., the origin and structure of the lesson; etc.)

3. What do you hope to accomplish in this lesson?

4. What mathematical ideas are embedded in this lesson?

5. What have you and your students been working on prior to this lesson?

6. How does this lesson fit into your overall goals for the year?

7. Are there students who have special issues in the class?

Learning & Pedagogy Observation Guide

Students		
Focus Question	Conjectures	Evidence from Classroom
What kinds of mathematical sense making are students doing?		
• What are the partially formed ideas that students are working with?		
• What ideas are the students particularly engaged in thinking about?		
• What ideas seem to be well understood?		
• What ideas seem to be poorly understood?		
What mathematical ideas seem to be confusing to students?		
• What might the root of the confusion be?		
• What new thinking does the confusion lead to?		
How are students developing their mathematical ideas over time?		
• In what instances do students refer to prior discussion, explorations, or findings?		
• Do students say anything about what they now understand or what they still need to learn?		
• What changes in thinking do you see that the children are not directly acknowledging?		

© Education Development Center, Inc.

Learning & Pedagogy Observation Guide

Teachers		
Evidence from Classroom	Conjectures	Focus Question
		How does the teacher work with the sense the children are making?
		• What opportunities does the teacher give students to share their thinking?
		• What does the teacher do to build on their ideas?
		• Does the teacher choose not to pursue certain ideas?
		How does the teacher work productively with students' confusions?
		• What does the teacher do to help students think through confusing ideas?
		Does the teacher seem to understand that students' ideas develop over time?
		• What does the teacher do to help students make connections to prior mathematical ideas?
		• How does the teacher work with students who are in different places with respect to any particular idea?
		• What expectations does the teacher have regarding students' levels of understanding based on the way the session ends?
		How does the teacher attend to all students?
		How does the teacher adjust her teaching based on the ideas that she hears from students?

Designing Packages Transcripts[2]

Transcript for Excerpt 1

G: Did we find all the methods? Because we took six times four times one equals twenty-four, four times three times two equals twenty-four, two times two times six equals twenty-four, twenty-four times one times one equals twenty-four, twelve times two times one equals twenty-four, eight times three times one equals twenty-four. But, like, we don't know if we can find any more.

T: You can't. You don't know if you can find any more? Osam, What were you doing over there?

O: I was using the multiplication chart to see what would times up to twenty-four.

T: How can you know if you've got all of them? All the possibilities? Do you have any ideas on that? Or think about it.

G. See if anything else equals twenty-four. Like, something times something equals twenty-four.

T. Okay, so that's something you guys can work on, okay?

Transcript for Excerpt 2

T: Look at all the charts. If you see anything interesting, raise your hand. Libby?

L: Like, everyone has, like, the same surface areas, just about, like I was looking at our group, and some other groups have the same numbers. And then, let's see, so with that group right there, for B, "which has like the most material and which uses the least material," that's exactly what we got—ninety-eight and fifty-two.

T: Libby, why is that, that each of the groups are having the same surface areas in their charts?

---

[2]From *Modeling Middle School Mathematics: The Connected Mathematics Project—Designing Packages, Grade 7*. Used with permission from L. Carey Bolster, Bolster Education, Inc. ©1992. For more information write to bolstereducation@yahoo.com

L: Because they have like the same length, width, and height that we used. Like, cause, like they probably had the same shape, and they sat it the same way we did, but, we sat it like this. But if they would have sat it like this, it would have been totally different.

T: What would have been totally different?

L: Like, if we would have sat it like this, the length would have only been two, but if someone else sat it like this, it would have been like twelve. . . . so that would change what they got for the surface area.

T: That would change what they got for the surface area.

L: Right.

T: Okay, Ginny, what do you think?

G: I think that she's wrong because the length and the width can be switched around but you still get the same surface area.

T: Okay, and why is that?

G: It's because, like, it's the same shape if you just like turn it around it'll still be the same shape, just different.

T: What do you think about that, Libby?

L: What we did at the beginning of class, when you showed us the Rice Krispie box, like when you turned it around and, like, it was still the same thing. So I was wrong, but . . . So even though if you do turn it like the surface area will stay the same.

T: Okay, the only thing that's different there is the way that we're having the box sit.. The area of the base is what we're changing. We can turn it all different ways, but what do we know about the surface area of this box? Erin?

E: It's always going to be the same thing. You just look at it differently.

Transcript for Excerpt 3

T: Are there any other patterns that anybody noticed?

T: Grace?

G: I noticed that as it became more like a cube the height increased.

T: As it became more like a cube the height got taller and taller. That's neat. Anyone have any idea why that happened? Libby?

L: Because when the cubes com . . . like, get smaller, they all like hook together. It gets taller because when you like compact 'em like it just makes it like that. And the surface area decreases because, when they're like all connected all the sides that are like facing out, there aren't as many because they're like hooked together but when they're in a straight line there's only one side that's actually touching the rest of them, but when they're all together there's like two or three sides.

T: When they're together like that, there's all those faces on the inside . . .

L: Yeah

T ... that never see the outside.

L: Yeah. Like this ... But in a straight line there's at least two or three showing.

E. I was going to say that for each cube in the straight line only two faces of that cube are not showing.

T: Neat. That's neat.

E. So it's really ... But if you pack it together like this, the least faces that any one cube has not showing is three, and that's the corner ones. All the others are more than three that aren't showing.

T: Great. These three on a corner piece ... On a corner piece we have these one, two, three faces showing. Three faces are inside. On a side piece, you've got one, two faces on the outside and then you've got how many on the inside?

E: Four.

T: Four. We lost 'em. Right?

E: Right

T: And that's how that surface area gets smaller and smaller. That's awesome.

[I would have never guessed that Evan and Libby would have found that as the cubes, as the structure became more and more cube-like that more and more of the faces on the inside became hidden. That totally blew my mind away when they did that.]

SESSION 8

Bringing It All Together: Distributing Supervisory Practices and Fostering Generative Growth

Increasingly, there is a recognition that generative learning is nurtured when teachers and administrators work together in learning communities, rather than working in isolation. Being part of a learning community allows for the exchange of understandings and perspectives that is central in promoting growth—what Brian Lord has termed "critical colleagueship."[3] When the entire school is part of the learning community, it becomes a place where not only students, but also adults in various teaching and administrative roles are learning. This combination of activity and reflection forms the foundation for generative learning.

In this course you have developed understandings of processes particular to mathematics teaching and learning so that you can engage in substantive discussions with teachers about the mathematics learning that is taking place in their classrooms.

In this session, you will consider the feasibility of redistributing the responsibility for classroom observation and supervision across different roles in the school. With such an approach, both the responsibility for and the learning that comes with conducting and discussing classroom observations can shared by teachers, staff developers, and administrators alike. In this session, you will consider these questions:

- What are the sources of your own authority and what is your capacity for influence as an instructional leader?

- How can concept of *generative learning* guide your supervisory work with teachers?

- How can classroom observation and supervision become a rich context for professional development for both teachers and yourself?

- How can the benefits of classroom observations afforded to both the teacher being observed and to the observer be distributed across roles in the school?

---

[3] Lord, B. (1994). Teachers' professional development: Critical colleagueship and the role of professional communities. In N. Cobb (Ed.), *The Future of Education: Perspectives on National Standards in America* (pp. 175–204). New York: College Entrance Examination Board.

Sources of Authority for Supervision

Read *Sources of Authority for Supervision,* by Sergiovanni and Starratt, (Reading 10) and think about the following question:

1. How do the ideas in this reading relate to what we have been re-examining in this course, particularly with respect to the two themes of the course, developing a new eye for what is important to look for in mathematics classrooms and rethinking the ways we talk with teachers about what we observe in their classrooms?

Then write about the following question:

2. What supervisory practices are currently in place in your school? Use Sergiovanni and Starrattt's description of sources of authority as a framework for examining the kinds of authority you call upon to support these practices.

Finally, reread what you wrote in Session 1 of this course about what you need to know in order to be a good observer and what you gain from conducting classroom observations. Amplify and extend your response in order to reflect your current understandings.

Sources of Authority for Supervision*

Thomas Sergiovanni and Robert Starratt

The Issue of Change

Change does not come about easily and is very difficult to mandate from the top down or from the outside. Mandated change requires more checking and monitoring to sustain than is possible to provide. Further, for change to have meaning and effect it must change not only the way things look but also the way things work. And finally, too often efforts to change are directed only toward doing the same things better. Change that counts, by contrast, is typically that which alters basic issues of schooling such as goals, values, beliefs, working arrangements, and the distribution of power and authority. This kind of change requires more than just tinkering with the existing school culture.

Teacher evaluation provides a good example. In recent years those interested in improving teaching and learning have frequently relied upon state-mandated evaluation systems that feature checklists to track certain observed teaching behaviors. Yet the record to date suggests that such mandates have not made much of a difference in improving the quality of teaching and learning. For example, teachers may conform to the mandated system only when under scrutiny. Since constant checking and monitoring are not possible, such systems soon become time-consuming and expensive. Using such monitoring, the school can supply evaluation data that suggest teachers are acting in required ways and that point to an array of new policies and procedures that are in place in the school as further evidence. Nevertheless, the teaching and learning process continues as it did before.

In the case above, changes in teaching practice are superficial rather than real. One reason for this superficiality is that the source of authority chosen for implementing changes is too limited. Present supervisory practices emerge from a particular pattern of authority. Changing these practices means changing the authority base for supervision. If, for example, teaching is viewed as a profession within which practice is based on research, the wisdom of experience, careful analysis of the situation at hand, and commitment to professional virtue, then the sources of authority for what teachers do would be internal, knowledge oriented,

---

*from *Supervision: A Redefinition* by T. Sergiovanni and R. Starratt. Boston, MA: McGraw Hill. Copyright ©1993. Reprinted by permission of the publisher. All rights reserved.

and norm based. If, by contrast, teaching continues to be viewed as a technical field within which practice is based on set routines and practitioners are in need of constant direction and monitoring, then the sources of authority for what teachers do are external, process oriented, and management based.

The Sources of Authority

Supervisory policies and practices can be based on one or a combination of five broad sources of authority:

Bureaucratic, in the form of legal and organizational mandates, rules, regulations, job descriptions, and expectations. When supervisory policies and practices are based on bureaucratic authority, teachers are expected to respond appropriately or face the consequences.

Personal, in the form of interpersonal leadership, motivational technology, and human relations skills. When supervisory policies and practices are based on personal authority, teachers are expected to respond to the supervisor's personality, to the pleasant environment provided, and to incentives for positive behavior. Personal authority is enhanced by learning how to apply insights from psychology and human and organizational behavior.

Technical-rational, in the form of evidence derived from logic and scientific research in education. When supervisory policies and practices are based on the authority of technical rationality, teachers are expected to respond according to what is considered the truth. Research, for example, tells teachers what to do rather than informs the decision they make about what to do.

Professional, in the form of experience, knowledge of the craft, and personal expertise. When supervisory policies and practices are based on professional authority, teachers are expected to respond to common socialization, accepted tenets of practice, and internalized expertise. Research, in this case, does not tell teachers what to do but informs the decision that they make about what to do.

Moral, in the form of obligations and duties derived from widely shared values, ideas, and ideals. When supervisory policies and practices are based on moral authority, teachers are expected to respond to shared commitments and felt interdependence.[1]

Each of the five sources of authority is legitimate and should be used, but the impact on teachers and on the teaching and learning process depends on which source or combination of sources is prime.

In the sections that follow we take a closer look at each of the five sources of authority, examining the assumptions underlying each one when it is used as the prime source, the supervision strategies suggested by each, and the impact each one has on the work of teachers and on the teaching and learning process.

---

[1]The discussion of sources of authority for supervision in this chapter is based on Thomas J. Sergiovanni, "Moral Authority and the Regeneration of Supervision," in Carl Glickman (ed). *Supervision in Transition.* The 1992 Yearbook of the Association for Supervision and Curriculum Development, Alexandra, VA., 1992; and "The Sources of Authority for Leadership" in Thomas J. Sergiovanni, *The Moral Dimensions in Leadership,* San Francisco: Jossey-Bass, 1992, chap. 3.

Bureaucratic Authority

As suggested above, bureaucratic authority relies heavily on hierarchy, rules and regulations, mandates, and clearly communicated role expectations as a way to provide teachers with a script to follow. Teachers, in turn, are expected to comply with this script or face consequences. There may be a place for this source of authority even in the most progressive of enterprises but when this source of authority is prime, the following assumptions are made:

- Teachers are subordinates in a hierarchically arranged system.
- Supervisors are trustworthy, but you can't trust subordinates very much.
- The goals and interests of teachers and those of supervisors are not the same; thus, supervisors must be watchful.
- Hierarchy equals expertise; thus, supervisors know more about everything than do teachers.
- External accountability works best.

With these assumptions in place it becomes important for supervisors to provide teachers with prescriptions for what, when, and how to teach, and for governing other aspects of their school lives. These are provided in the form of expectations. Supervisors then practice a policy of "expect and inspect" to ensure compliance with these prescriptions. Heavy reliance is placed on predetermined standards to which teachers must measure up. Since teachers often will not know how to do what needs to be done, it is important for supervisors to identify their needs and then to "in-service" them in some way. Directly supervising and closely monitoring the work of teachers is key in order to ensure continued compliance with prescriptions and expectations. To the extent possible, it is also a good idea to figure out how to motivate teachers and encourage them to change in ways that conform with the system.

The consequences of relying on bureaucratic authority in supervision have been carefully documented in the literature. Without proper monitoring, teachers wind up being loosely connected to bureaucratic systems, complying only when they have to.[2] When monitoring is effective in enforcing compliance, teachers respond as technicians who execute predetermined scripts and whose performance is narrowed. They become, to use the jargon, "deskilled."[3] When teachers are not able to use their talents fully and are caught in the grind of routine, they become separated from their work, viewing teaching as a job rather than a vocation, and treating students as cases rather than persons.

---

[2]See, for example, Karl Weick, "Educational Organizations as Loosely Coupled Systems," *Administrative Science Quarterly*, vol. 21, no. 2 (1976), pp. 1–19; and Thomas J. Sergiovanni, "Biting the Bullet: Rescinding the Texas Teacher Appraisal System," *Teachers Education and Practice*, vol. 6, no. 2 (Fall/Winter 1990–1991), 89–93.

[3]See, for example, Arthur E. Wise, *Legislated Learning: The Bureaucratization of the American Classroom*. Berkley, Calif.: University of California Press, 1979; Susan Rosenholtz, *Teachers' Workplace: The Social Organization of Schools*. New York: Longman, 1989; and Linda McNeil, *Contradictions of Control: School Structure and School Knowledge*. New York: Routledge & Kegan Paul, 1986.

Readers probably will have little difficulty accepting the assertion that supervision based primarily on bureaucratic authority is not a good idea. The validity of most of the assumptions underlying this source of authority are suspect. Few, for example, believe that teachers as a group are not trustworthy and do not share the same goals and interests about schooling as do their supervisors. Even fewer would accept the idea that hierarchy equals expertise. Less contested, perhaps, would be the assumptions that teachers are subordinates in a hierarchically arranged system and that external monitoring works best. Supervision today relies heavily on predetermined standards. Because of this, supervisors need to spend a good deal of time trying to figure out strategies for motivating teachers and encouraging them to change. Supervision becomes a direct, intense, and often exhausting activity.

Personal Authority

Personal authority is based on the supervisor's leadership expertise in using motivational techniques and in practicing other interpersonal skills. It is assumed that as a result of this leadership, teachers will want to comply with the supervisor's wishes. When human relations skills become the prime source of authority, the following assumptions are made:

- The goals and interests of teachers and supervisors are not the same. As a result each must barter with the other so that both get what they want by giving something that the other party wants.

- Teachers have needs; if these needs are met, the work gets done as required in exchange.

- Congenial relationships and harmonious interpersonal climates make teachers content, easier to work with, and more apt to cooperate.

- Supervisors must be experts at reading the needs of teachers and handling people in order to barter successfully for increased compliance and performance.

These assumptions lead to a supervisory practice that relies heavily on "expect and reward" and "what gets rewarded gets done." Emphasis is also given to developing a school climate characterized by a high degree of congeniality among teachers and between teachers and supervisors. Often personal authority is used in combination with bureaucratic and technical-rational authority. When this is the case very few of the things that the supervisor wants from teachers are negotiable. The idea is to obtain compliance by trading psychological payoffs of one sort or another.

Personal authority is also important to the practice of human resource leadership. In this case, however, as suggested in Chapter 1, it takes a slightly different twist. The emphasis is less on meeting teachers' social needs and more on providing the *conditions of work* that allow people to meet needs for achievement, challenge, responsibility, autonomy, and esteem—the presumed basis for finding deep psychological fulfillment in one's job.

The typical reaction of teachers to personal authority, particularly when connected to human relations supervision, is to respond as required when rewards are available but not otherwise. Teachers become involved in their work for calculated reasons, and their performance becomes increasingly narrowed. When the emphasis is on psychological fulfillment that comes from the work itself (emphasizing that in work, for example) rather than the supervisor's skilled interpersonal behavior, teachers become more intrinsically motivated and thus less susceptible to calculated involvement and narrowing of performance. But in today's supervision this emphasis remains the exception rather than the rule.

Suggesting that using personal authority and the psychological theories that inform this authority as the basis for supervisory practice may be overdone and may have negative consequences for teachers and students is likely to raise a few eyebrows. Most supervisors, for example, tend to consider knowledge and skill about how to motivate teachers, how to apply the correct leadership style, how to boost morale, and how to engineer the right interpersonal climate as representing the heart of their work. It is for many supervisors the "core technology" of their profession.

We do not challenge the importance of psychologically based supervision and leadership. Indeed, we argued for its importance in earlier editions. We do question, however, whether it should continue to enjoy the prominence that it does. Our position is that personal, bureaucratic, and technical-rational sources of authority should do no more than provide support for a supervisory practice that relies on professional and moral authority. The reasons, we argue here and elsewhere in this book, are that psychologically based leadership and supervision cannot tap the full range and depth of human capacity and will. This source of authority cannot elicit the kind of motivated and spirited response from teachers that will allow schools to work well.

Another reason for our concern with the overuse of personal authority is that there are practically and morally better reasons for teachers and others to be involved in the work of the school than those related to matters of the leader's personality and interpersonal skills. Haller and Strike, for example, believe that building one's expertise around interpersonal themes raises important ethical questions. In their words:

> We find this an inadequate view of the administrative role . . . its first deficiency is that it makes administrative success depend on characteristics that tend to be both intangible and unalterable. One person's dynamic leader is another's tyrant. What one person sees as a democratic style, another will see as the generation of time-wasting committee work. . . . Our basic concern with this view . . . is that it makes the administrative role one of form, not content. Being a successful administrator depends not on the adequacy of one's view, not on the educational policies that one adopts and how reasonable they are, and not on how successful one is in communicating those reasons to others. Success depends on personality and style, or on carefully chosen ways of inducing others to contribute to the organization. It is not what one wants to do and why that is important; it is who one is and how one does things that counts. We find such a view offensive. It is incompatible with the values of autonomy, reason and democracy, which we see among the central commitments of our society and educational system. Of course educational administrators must be leaders, but let them lead by reason and persuasion, not by forces of personality.[4]

---

[4]Emil J. Haller and Kenneth A. Strike, *An Introduction to Educational Administration: Social, Legal and Ethical Perspectives.* New York: Longman, 1986, p. 326.

Carl Glickman raises still other doubts about the desirability of basing supervisory practice on psychological authority. He believes that such leadership creates dependency among followers.[5]

The perspectives of Haller and Strike and Glickman raise a nagging set of questions. Why should teachers follow the lead of their supervisors? Is it because supervisors know how to manipulate effectively? Is it because supervisors can meet the psychological needs of teachers? Is it because supervisors are charming and fun to be with? Or is it because supervisors have something to say that makes sense? Or because supervisors have thoughts that point teachers in a direction that captures their imagination? Or because supervisors speak from a set of ideas, values, and conceptions that they believe are good for teachers, for students, and for the school? These questions raise yet another question: Do supervisors want to base their practice on glitz or on substance? Choosing glitz not only raises moral questions but also encourages a vacuous form of leadership and supervisory practice. It can lead to what Abraham Zaleznik refers to as the "managerial mystique," the substitution of process for substance.[6]

Technical-Rational Authority

Technical-rational authority relies heavily on evidence that is defined by logic and scientific research. Teachers are expected to comply with prescriptions based on this source of authority in light of what is considered to be truth.

When technical rationality becomes the primary source of authority for supervisory practice, the following assumptions are made:

- Supervision and teaching are applied sciences.
- Scientific knowledge is superordinate to practice.
- Teachers are skilled technicians.
- Values, preferences, and beliefs are subjective and ephemeral; facts and objective evidence are what matters.

When technical-rational authority is established, then the supervisory strategy is to use research to identify what is best teaching practice and what is best supervisory practice. Once this is known, the work of teaching and supervision is standardized to reflect the best way. The next step is to in-service teachers in the best way. For the system to work smoothly, it is best if teachers willingly conform to what the research says ought to be done and how it ought to be done. Thus it is important to figure out how to motivate teachers and encourage them to change willingly.

When technical-rational authority is the primary source, the impact on teachers is similar to that of bureaucratic authority. Teachers are less likely to conform to what research says and more likely to act according to their beliefs. When forced to conform, they are likely to respond as technicians executing predetermined steps,

---

[5] Carl D. Glickman, "Right Question, Wrong Extrapolation: A Response to Duffey's 'Supervising for Results,'" *Journal of Curriculum and Supervision,* vol. 6, no. 1 (Fall 1990), pp. 39–40.

[6] Abraham Zaleznik, *The Managerial Mystique Restoring Leadership in Business.* New York: Harper & Row, 1989.

and their performance becomes increasingly narrowed. When technical-rational authority is used in culmination with personal authority, teachers tend to conform as long as they are being rewarded.

If criticism of a supervisory practice based on personal authority raises concerns, then suggesting that primary use of technical-rational authority is dysfunctional is also likely to raise concerns. We live, after all, in a technical-rational society where that which is considered scientific is prized. Because of this deference to science, the above beliefs and their related practices are likely to receive ready acceptance. "Supervision and teaching are applied sciences" has a nice ring to it, and using research to identify one best practice seems quite reasonable. But teaching and learning are too complex to be captured so simply. In the real world of teaching none of the assumptions hold up very well, and the related practices portray an unrealistic view of teaching and supervision.

For example, there is a growing sense among researchers, teachers, and policy analysts that the context for teaching practice is too idiosyncratic, nonlinear, and loosely connected for simplistic conceptions of teaching to apply.[7] Teaching cannot be standardized. Teachers, like other professionals, cannot be effective when following scripts. Instead they need to *create knowledge in use* as they practice, becoming skilled surfers who ride the wave of teaching as it uncurls.[8] This ability requires a higher level of reflection, understanding, and skill than that required by the practice of technical rationality—a theme to be further developed in Parts Three and Four.

The authority of technical rationality for supervisory practice does share some similarities with the authority of professionalism. Both, for example, rely on expertise. But the authority of technical rationality presumes that scientific knowledge is the only source of expertise. Further, this knowledge exists separate from the context of teaching. The job of the teacher is simply to apply this knowledge in practice. In other words, the teacher is *subordinate* to the knowledge base of teaching.

Professional Authority

Professional authority presumes that the expertise of teachers counts most. Teachers, as is the case with other professionals, are always *superordinate* to the knowledge base that supports their practice. Professionals view knowledge as something that informs but does not prescribe practice.[9] What counts as well is the ability of teachers to

---

[7]See, for example, Linda Darling-Hammond, Arthur E. Wise, and S. R. Pease, "Teacher Evaluations in an Organizational Context: A Review of Literature," *Review of Educational Research,* vol. 53, no. 3 (1983), pp. 285–328; Thomas J. Sergiovanni, "The Metaphorical Use of Theories and Models in Supervision and Teaching: Building a Science," *Journal of Curriculum and Supervision,* vol. 2, no. 3 (Spring 1987), pp. 221–232; Lee S. Shulman, "A Union of Insufficiencies: Strategies for Teacher Assessment in a Period of Educational Reform," *Educational Leadership,* vol. 16, no. 3 (1988), pp. 36–41; and Michael Huberman, "The Social Context of Instruction in Schools." Paper presented at American Educational Research Association Annual Meeting, Boston, April 1990.

[8]Thomas J. Sergiovanni, "Will We Ever Have a TRUE Profession?" *Educational Leadership,* vol. 44, no. 8 (1987), pp. 44–51.

[9]Thomas J. Sergiovanni, "The Metaphorical Use of Theories and Models in Supervision: Building a Science," *Journal of Curriculum and Supervision,* vol. 2, no. 3 (1987), pp. 221–232.

make judgments based on the specifics of the situations they face. They must decide what is appropriate. They must decide what is right and good. They must, in sum, create professional knowledge in use as they practice.

Professional authority is based on the informed knowledge of the craft of teaching and on the personal expertise of teachers. Teachers respond in part to this expertise and in part to internalized professional values, to accepted tenets of practice that define what it means to be a teacher.

When professional authority becomes the primary source for supervisory practice, the following assumptions are made:

- Situations are idiosyncratic; thus, no one best way to practice exists.
- "Scientific knowledge" and "professional knowledge" are different; professional knowledge is created as teachers practice.
- The purpose of "scientific knowledge" is to inform, not prescribe, the practice of teachers and supervisors.
- Professional authority is not external but is exercised within the teaching context and from within the teacher.
- Authority in context comes from the teacher's training and experience.
- Authority from within comes from the teacher's professional socialization and internalized knowledge and values.

Supervisory practice that is based primarily on professional authority seeks to promote a dialogue among teachers that makes explicit professional values and accepted tenets of practice. These are then translated into professional practice standards. With standards acknowledged, teachers are then provided with as much discretion as they want and need. When professional authority is fully developed, teachers will hold each other accountable in meeting these practice standards with accountability internalized. The job of the supervisor is to provide wide assistance, support, and professional development opportunities. Teachers respond to professional norms, and their performance becomes more expansive.

Though it is common to refer to teaching as a profession, not much attention has been given to the nature of professional authority. When the idea does receive attention, the emphasis is on the expertise of teachers. Building teacher expertise is a long-term proposition. In the meantime much can be done to advance another aspect of professionalism—*professional virtue*. Professional virtue speaks to the norms that define what it means to be a professional. Once established, professional norms take on moral attributes. When professional norms are combined with norms derived from shared community values, moral authority can become a prime basis for supervisory practice.

Moral Authority

Moral authority is derived from the obligations and duties that teachers feel as a result of their connection to widely shared community values, ideas, and ideals. When moral authority is in place, teachers respond to shared commitments and interdependence by becoming self-managing.

When moral authority becomes the primary source for supervisory practice, schools can become transformed from organizations into communities. Communities are defined by their center of shared values, beliefs, and commitments. In communities, what is considered right and good is as important as what works and what is effective; teachers are motivated as much by emotion and belief as they are by self-interest; collegiality is understood as a form of professional virtue.

In communities, supervisors direct their efforts toward identifying and making explicit shared values and beliefs. These values and beliefs are then transformed into informal norms that govern behavior. With these in place it becomes possible to promote collegiality as something that is internally felt and that derives from morally driven interdependence. Supervisors can rely less on external controls and more on the ability of teachers as community members to respond to felt duties and obligations. The school community's informal norm system is used to enforce professional and community values. Norms and values, whether derived from professional authority or moral authority, become substitutes for direct supervision as teachers become increasingly self-managing. The five sources of authority for supervision with consequences for practice are summarized in Table 3–1.

Table 3–1

The Sources of Authority for Supervisory Policy and Practice

Source	Assumptions When Use of This Source Is Prime	Leadership/Supervisory Strategy	Consequences
Bureaucratic authority Hierarchy Rules and regulations Mandates Role expectation Teachers are expected to comply or or face consequences.	–Teachers are subordinates in a hierarchically arranged system. –Supervisors are trustworthy, but you cannot trust subordinates very much. –Goals and interests of teachers and supervisors are not the same; thus, supervisors must be watchful. –Hierarchy equals expertise; thus, supervisors know more than teachers. –External accountability works best.	–"Expect and inspect" is the overarching rule. –Rely on predetermined standards to which teachers must measure up. –Identify teachers' needs and "in-service" them. –Directly supervise and closely monitor the work of teachers to ensure compliance. –Figure out how to motivate teachers and get them to change.	–With proper monitoring teachers respond as technicians in executing predetermined scripts. Their performance is narrowed.
Personal authority Motivation technology Interpersonal skills Human relations leadership Teachers will want to comply because of the congenial climate provided and to reap rewards offered in exchange.	–The goals and interests of teachers and supervisors are not the same but can be bartered so that each gets what they want. –Teachers have needs; if those needs are met, the work gets done as required in exchange. –Congenial relationships and harmonious interpersonal climates make teachers content, easier to work with and more apt to cooperate. –Supervisors must be experts at reading needs and handling people in order to successfully barter for increased compliance and performance.	–Develop a school climate characterized by congeniality among teachers and between teachers and supervisors. –"Expect and reward." –"What gets rewarded gets done." –Use personal authority in combination with bureaucractic and technical authority.	–Teachers respond as required when rewards are available but not otherwise. Their involvement is calculated, and performance is narrowed.
The authority of technical rationality Evidence defined by logic and scientific research Teachers are required to comply in light of what is considered to be truth.	–Supervision and teaching are applied sciences. –Knowledge of research is privileged. –Scientific knowledge is superordinate to practice. –Teachers are skilled technicians. –Values, preferences, and beliefs don't count, but facts and objective evidence do.	–Use research to identify one best practice. –Standardize the work of teaching to reflect the best way. –"In-service" teachers in the best way. –Monitor the process to ensure compliance. –Figure out ways to motivate teachers and get them to change.	–With proper monitoring,teachers respond as technicians in executing predetermined scripts. Performance is narrowed.

Table 3–1 continued

Professional authority Informed knowledge of craft and personal expertise. Teachers respond on basis of common socialization, professional values, accepted tenets of practice, and internalized expertness.	–Situations are idiosyncratic; thus, no best way exists. –"Scientific" knowledge and "professional" knowledge are different; professional is created in use as teachers practice. –The purpose of "scientific" knowledge is to inform, not prescribe, practice. –Authority cannot be external but comes from the context itself and from within the teacher. –Authority from context comes from training and experience. –Authority from within comes from socialization and internalized values.	–Promote a dialogue among teachers that makes explicit professional values and accepted tenets of practice. –Translate above into professional practice standards. –Provide teachers with as much discretion as they want and need. –Require teachers to hold each other accountable in meeting practice standards. –Make available assistance, support, and professional development opportunities.	–Teachers respond to professional norms and those require little monitoring. Their performance is expansive.
Moral authority Full obligations and duties derived from widely shared community values, ideas, and ideals. Teachers respond to shared commitments and felt interdependence.	–Schools are professional learning communities. –Communities are defined by their center of shared values, beliefs, and communities. –In communities: –what is considered right and good is as important as what works and what is effective. –People are motivated as much by emotion and beliefs as by self-interest. –Collegiality is a professional virtue.	–Identify and make explicit the values and beliefs that define the center of the school as community. –Translate the above into informal norms that govern behavior. –Promote collegiality as internally felt and morally driven interdependence. –Rely on ability of community members to respond to duties and obligations. –Rely on the community's informal norm system to enforce professional and community values.	–Teachers respond to community values for moral reasons. Their performance is expansive and sustained.

Supervision II

In this section some of the basic assumptions and underlying operating principles that differentiate Supervision I from Supervision II are provided. Supervision II combines Theory Y from human resources supervision with the belief that people are morally responsive and able to sacrifice self-interests for the right reasons. Supervision 1, by contrast, relies heavily on the assumptions of Theory X (scientific management) and Theory X soft (human relations), as described earlier.

In Supervision I a great deal of emphasis is given to understanding, researching, and improving supervisory behavior. As will be discussed in Part Two, Supervision II emphasizes action. Behavior is very different from action. Behavior is what we do on the surface. Action, by contrast, implies intentionality, free choice, value seeking, and altruism.

Because of this difference between behavior and action, Supervision II focuses more on interpretation. There is concern not only for the way things look but also for what things mean. The metaphors "phonetics" and "semantics" can help this distinction.[10] Tending to supervisory and teaching behaviors as they appear on the surface is an example of the phonetic view. It does not matter so much whether the supervisor is involved in leading, coaching, managing, evaluating, administering, or teaching. If the emphasis in these activities is on "the looks and sounds" of behavior, on the form or shape that this behavior takes as opposed to what the behavior means to teachers and students, the view is phonetic.

Identical behaviors can have different meanings as contexts change and as different people are involved. For example, a supervisor may walk though the classrooms of several teachers on a regular basis, making it a practice to comment to teachers about what is happening and to share her or his impressions. For supervisor A, teachers may consider this behavior inspectorial or controlling and view this supervisor as one who is closely monitoring what they do. For supervisor C this same behavior may be considered symbolic of the interest and support that the supervisor provides to teachers. In this case supervisory behavior is interpreted as being caring and helpful. At one level, the phonetic, the behavior is the same for both supervisors. At another level the semantic, the behavior carries different meanings. When concerned with different interpretations and meanings, one is tending to the semantic aspects of supervision.

Motivation in Supervision I focuses primarily on "what gets rewarded gets done." Supervision II is based on "what is rewarding gets done" and "what is believed to be right and good gets done." These latter emphases, as will be discussed in Chapter 5, not only reflect more completely what is important to teachers at work but also result in less emphasis on direct control-oriented supervision. When motivated by intrinsically satisfying and meaningful action, teachers become self-managing.

In Supervision I it is assumed that supervisors and teachers make "rational" decisions on the basis of self-interest and as isolated individuals. Supervision II recognizes the importance of emotions and values in making decisions and the

---

[10]Sergiovanni, *Educational Leadership,* op. cit.

capacity for humans consistently to sacrifice self-interest as a result. Further, Supervision II recognized that our connections to other people to a great extent determine what we think, what we believe, and the decisions that we make. These two themes will be explored further in Chapters 5 and 6.

Supervision I takes place in the context of hierarchically differentiated roles. "Supervision" and "designated supervisor" go together. Supervision is something that supervisors do to teachers. A teacher studies to becomes a supervisor, becomes licensed, and thus is allowed to practice supervision. Supervision, in other words, is a formal and institutionalized process linked to the school's organizational structure.

Supervision II views supervision as a process and function that is hierarchically independent and role-free. It may be linked to hierarchy and role, but it need not be. Supervision, in other words, is not necessarily shaped by organizational structure and is not necessarily legitimized by credentials. Instead, it is a set of ideals and skills that can be translated into processes that can help teachers and help schools function more effectively. Supervision is something that not only principals and hierarchically designated supervisors do but also teachers and others. Thus, such ideas as collegial supervision, mentoring supervision, cooperative supervision, and informal supervision are important in Supervision II. In a sense, these processes are often in place in schools anyway. Teachers respect each other's craft knowledge and depend on each other for help. One of the purposes of Supervision II, therefore, is to deinstitutionalize institutional supervision and to formalize the informal supervision among teachers that now takes place in schools.

Models of Control

Supervision is a process designed to help teachers and supervisors learn more about their practice, to be better able to use their knowledge and skills to serve parents and schools, and to make the school a more effective learning community. For these goals to be realized, a degree of control over events is necessary, and in this sense supervision is about control. But it makes a difference how control is expressed in schools. The wrong kind of control can cause problems and lead to negative consequences despite the best of intentions.

Control is understood differently in Supervision I and II. For example, the management theorist Henry Mintzberg proposed that the work of others can be controlled by providing direct and close supervision of what people are doing; by standardizing the work that needs to be done, then fitting people into the work system so that they are forced to follow the approved script; by standardizing outputs and then evaluating to be sure that output specifications have been met; by socializing people through the use of norms of one sort or another; and by arranging work circumstances and norms in a way that people feel a need to be interdependent.[11] When referring to socializing people, Mintzberg had in mind professional norms. The norms that come from common purposes and shared values provide still another control strategy that can be added to Mintzberg's list.[12]

---

[11]Henry Mintzberg, *The Structure of Organizations.* New York: Wiley, 1979.

[12]Karl Weick, "The Concept of Loose Coupling: An Assessment," *Organizational Theory Dialogue,* (December 1986); and Thomas J. Peters and Robert H. Waterman, *In Search of Excellence.* New York: Harper & Row, 1982.

In Chapter 1 supervisor A relied heavily on providing direct supervision, standardizing the work, and standardizing outputs as the preferred ways to control what teachers were doing and how. By contrast, supervisor D recommended relying on the process of socialization, building interdependencies, and developing purposes and shared values. Which of the two views makes the most sense? In part, the answer to this question depends upon the degree of complexity of the work to be supervised.

Fostering professional socialization, developing purposes and shared values, and building natural interdependencies among teachers are unique in that they are able to provide a kind of normative power that encourages people to meet their commitments. Once in place, the three become substitutes for traditional supervision, since teachers tend to respond from within, becoming self-managing. Since these strategies do not require direct supervision or the scripting of work, they are better matched to the complex behaviors that are required for teaching and learning to take place successfully. These are the strategies that provide the framework for control in Supervision II. The three are much more difficult to implement when schools are viewed as formal organizations. Such organizations tend to nurture control systems that rely on direct supervision, standardized work, and standardized outputs. In the next chapter we propose that one way to improve schools is to change the way they are understood, transforming them from organizations into communities.

The Importance of Capacity Building

From the discussion so far, it appears that Supervision II is the preferred method and no room exists for practices associated with Supervision I. Similarly, bureaucratic and personal authority should be abandoned for leadership and supervision based on professional and moral authority. But the reality is that ideas from both views of supervision have a place in a unified and contextually oriented practice. Most teachers have a natural inclination to respond to Supervision II. As this approach comes to dominate supervisory practice, schools will very likely be better places for teachers and students. But some teachers will not be ready to respond. Others may want to respond but lack the necessary knowledge and skill to function in this new environment. Furthermore, in many places the existing culture of teaching does not encourage the practices associated with Supervision II. What should supervisors do?

We believe the answer is to start with where teachers are now and to begin the struggle to change the existing norms in schools so that Supervision II becomes acceptable. Of particular importance in this struggle will be the emphasis supervisors give to capacity building. Take teacher leadership, for example. Before we can expect teachers to accept fuller responsibility for providing leadership, they must be encouraged to do so and they must know how to provide this leadership. Both of these goals are best achieved when teachers are members of a support network that strives to become a community of leaders. For teachers to be developers and supervisors of classroom learning communities, they must be part of a learning community themselves. Capacity building and changing school norms are what Michael Fullan describes as "reculturing."[13] Whatever is the focus of supervisory work, if Supervision II is to be the framework for embodying that focus, then reculturing will be at the heart of supervisory work.

---

[13] Michael Fullan, *Change Forces*. London: Falmer Press, 1993.

Lenses on Learning

Observation Guide

	Students	
Focus Question	**Conjectures**	**Evidence from Classroom**
Math Content		
• What mathematical ideas are embedded in the lesson?		
• What makes this worthwhile mathematics?		
Learning		
• What kinds of mathematics sense-making are students doing?		
• What mathematical ideas seem to be confusing to students?		
• In what ways can you see that the students are developing their mathematical ideas over time?		
Intellectual Community		
• How are students showing respect for one another's ideas?		
• How do students use each other as resources as they make sense of mathematical ideas?		
• What evidence beyond raised hands do you have that students are engaged?		

© Education Development Center, Inc.

Lenses on Learning

Observation Guide

Teachers		
Evidence from Classroom	Conjectures	Focus Question
		Knowledge of Content
		• What does the teacher seem to understand about the mathematics?
		• What is the teacher's long-term mathematical agenda?
		• What does the teacher seem to understand about the development of children's ideas in this topic?
		Pedagogy
		• How does the teacher work with the sense the children are making?
		• How does the teacher work productively with students' confusion?
		• How does the teacher attend to all students?
		• How does the teacher adjust her teaching based on the ideas she hears from students?
		Facilitating Intellectual Community
		• How does the teacher support students in showing respect for one another's ideas?
		• How does the teacher set the tone so students see each other as resources for mathematical thinking?
		• What interventions does the teacher make to ensure that students' engagement has a focus on mathematical ideas?

© Education Development Center, Inc.

Intellectual Community Observation Guide

	Students	
Focus Question	Conjectures	Evidence from Classroom
How are students showing respect for one another's ideas?		
• What are students doing to show that they are listening to other students?		
• How willing are students to share their ideas even if they know that they aren't correct?		
• How attentive are students to one another's ideas?		
How do students use each other as resources as they make sense of mathematical ideas?		
• Are they building on each other's mathematical ideas?		
• Are they asking each other questions related to mathematical ideas?		
What evidence beyond raised hands do you have that students are engaged?		

© Education Development Center, Inc.

Lenses on Learning

Intellectual Community Observation Guide

Teachers		
Evidence from Classroom	Conjectures	Focus Question
		How does the teacher support students in showing respect for one another's ideas?
		• How does the teacher set and maintain norms of interaction for discourse?

		• How does the teacher support such practices as:
		  • attentive listening
		  • question-posing about mathematical ideas
		  • provisional thinking

		How does the teacher set the tone for students to see each other as resources for mathematical thinking?
		• Does the teacher invite students' tentative thinking?
		• Does the teacher invite students to build on one another's ideas?
		What interventions does the teacher make to ensure that students' engagement has a focus on mathematical ideas?

© Education Development Center, Inc.

Lenses on Learning

Math Content Observation Guide

	Students	
Focus Question	Conjectures	Evidence from Classroom
What mathematical ideas are embedded in the lesson?		
• What is the topic?		
• What are the ideas within this topic that are being explored?		
• What specific ideas are being explored by different students or groups of students?		
What makes this worthwhile mathematics?		
• What important mathematical ideas are involved?		
• What is the relationship between doing procedures and exploring ideas in this mathematics?		
• What kinds of mathematical thinking are taking place (conjectures, proofs, revising, generalizing, etc)?		

© Education Development Center, Inc.

Lenses on Learning

Math Content Observation Guide

Teachers		
Evidence from Classroom	**Conjectures**	**Focus Question**
		What does the teacher seem to understand about the mathematics?
		• What aspects of the mathematics does the teacher seem to know well and in what areas does he or she still seem to need to deepen her knowledge?
		• Is the teacher responding to the mathematics in students' mathematical thinking?
		• How does the teacher show that he or she understands enough about the mathematics to test the boundaries of students' understanding?
		What does the teacher seem to understand about the development of children's ideas in this topic?
		• How does the teacher show that he or she understands students' thinking?
		• What follow-up questions does the teacher ask to probe the robustness of students' understanding?
		• In what ways does the teacher help them extend or deepen their thinking?
		What seems to be the teacher's long-term mathematical agenda?
		• What mathematical ideas is the teacher probing?
		• What does the teacher do to bring into focus the long-term importance of these mathematical ideas?

© Education Development Center, Inc.

Lenses on Learning

Learning & Pedagogy Observation Guide

	Students	
Focus Question	Conjectures	Evidence from Classroom
What kinds of mathematical sense making are students doing?		
• What are the partially formed ideas that students are working with?		
• What ideas are the students particularly engaged in thinking about?		
• What ideas seem to be well understood?		
• What ideas seem to be poorly understood?		
What mathematical ideas seem to be confusing to students?		
• What might the root of the confusion be?		
• What new thinking does the confusion lead to?		
How are students developing their mathematical ideas over time?		
• In what instances do students refer to prior discussion, explorations, or findings?		
• Do students say anything about what they now understand or what they still need to learn?		
• What changes in thinking do you see that the children are not directly acknowledging?		

© Education Development Center, Inc.

154 ♦ Readings

Lenses on Learning

Learning & Pedagogy Observation Guide

Teachers		
Evidence from Classroom	Conjectures	Focus Question
		How does the teacher work with the sense the children are making?
		• What opportunities does the teacher give students to share their thinking?
		• What does the teacher do to build on their ideas?
		• Does the teacher choose not to pursue certain ideas?
		How does the teacher work productively with students' confusions?
		• What does the teacher do to help students think through confusing ideas?
		Does the teacher seem to understand that students' ideas develop over time?
		• What does the teacher do to help students make connections to prior mathematical ideas?
		• How does the teacher work with students who are in different places with respect to any particular idea?
		• What expectations does the teacher have regarding students' levels of understanding based on the way the session ends?
		How does the teacher attend to all students?
		How does the teacher adjust her teaching based on the ideas that she hears from students?

© Education Development Center, Inc.

Readings ♦ 155

RESOURCE LIST

Chapters and Articles

Ball, Deborah Loewenberg. (2000) Bridging Practices: Intertwining content and pedagogy in teaching and learning to teach. *Journal of Teacher Education,* 51(3), 241–47.

In this article, Ball examines how subject matter and pedagogy have been persistently divided in the conceptualization and curriculum of teacher education. The author discusses three problems that would have to be solved to bridge this gap and to prepare teachers who both know subject-matter content and can use it in making wise pedagogical choices. The problems include: identifying the content that matters for teaching, understanding how such knowledge needs to be held, and knowing what it takes to use such knowledge in practice.

Ball, Deborah Loewenberg. (1993). With an eye on the mathematical horizon: dilemmas of teaching elementary school mathematics. *The Elementary School Journal,* 93, (4).

Ball provides a powerful account of the challenges and dilemmas of developing a mathematics teaching practice that is both responsive to where children are in their mathematical understanding yet honors the long-established content of the discipline. Using examples from her own third-grade class, the author examines the kind of dilemmas that arise in three areas: the mathematical content itself, respecting children as mathematical thinkers, and creating and using community.

Ball, Deborah Loewenberg. (1992) Magical hopes: Manipulatives and the reform of math education. *American Educator, 16*(2), 14–18, 46–47.

Citing examples from her third grade mathematics classroom, the author describes problems stemming from the use of concrete objects or "manipulatives" in mathematics classrooms (e.g., fraction bars, base-10 blocks, and Popsicle sticks). The vignettes show the fallacy of assuming that students will automatically draw the conclusions that their teachers want simply by interacting with particular manipulatives.

Conference Board of the Mathematical Sciences. (2001) Recommendations for Elementary Teacher Preparation, Chapter 3 in *The Mathematical Education of Teachers.* Providence, RI: American Mathematical Society.

In this chapter, the authors describe what teachers need to understand in order to teach elementary mathematics (numbers and operations; algebra and functions; geometry and measurement; and data analysis, statistics, and probability). The chapter opens with a vignette from a third grade classroom about a teacher, who instead of taking her students through an algorithm step by step, probes their ideas in order to understand their thinking. The decisions the teacher makes and the thinking that underlies them are described, as are the understandings that the teacher needed to have to make these pedagogical decisions. How to transform a poorly prepared prospective elementary teacher into someone who can think mathematically is also addressed in this chapter.

Economopoulos, Karen. (1998) What comes next?: The mathematics of pattern in kindergarten. *Teaching Children Mathematics, 5*(4), 230–233.

This author discusses the purposes of developing facility with patterns in kindergarten. She illustrates how pattern activities, when well designed, can help children begin to develop their thinking about such complex mathematical ideas as predictability and consistency. She illustrates how such activities in the early grades help students connect with the mathematics of the later grades.

Goldenberg, E. Paul (2000) *Thinking (and talking) about technology in math classrooms.* Newton, MA: Education Development Center.

The author provides a framework for making decisions about technology use in mathematics classrooms. He points to research findings to stress that the value of technologies, from manipulatives used in the early grades to complex computer programs used in the higher grades, depends on how they are used. He suggests that dimensions such as genre of the technology, purpose of the lesson, nature of the thinking being asked of students, the role of the technology in the lesson, and prioritizing content and knowledgeable use of technology are important elements to consider.

Grant, Cathy Miles. (2000). Beyond just doing it: Making discerning decisions about using electronic graphing tools. *Learning & Leading with Technology, 27*(5) 14–17, 49.

Drawing on a project that incorporates both hand-drawn and computer-generated graphs to enrich second graders' understanding of data and what they represent, this author notes that the value of electronic graphing tools depends on teachers' recognition of the strengths and limitations of their use in the classroom.

Grant, C. M., and Lester, J. (2001) Mathematics supervision through a new lens, *Educational Leadership, 58*(5), 60–63.

Using illustrations from a *Lenses on Learning* seminar that took place in Western Massachusetts, these authors examine the kind of learning that can take place when school administrators focus on the mathematics content of the classroom. They describe the structure of the course and highlight how it helps participants to consider new ideas about mathematics, learning, and teaching, as well as their own needs for professional development within a standards-based context.

Hiebert, J., & Stigler, J. W. (2000). A proposal for improving classroom teaching: Lessons from the TIMSS video study. *Elementary School Journal, 101,* 3–20.

These authors discuss results from the TIMSS Video Study regarding U.S. teachers' view of reform. They presents the view of many teachers that they are changing the way they teach even though the core of their practice remains the same. The authors explore why this is the case and why it can be challenging to change such patterns of teaching in the United States. They conclude by offering suggestions for developing school-based, teacher-driven systems for improving teaching.

Lewis, C. & Tsuchida, I. (1998). A lesson is like a swiftly flowing river: Research lessons and the improvement of Japanese education. *American Educator,* Winter, 14–17 & 50–52.

These authors examine the structure of Japanese "research lessons," or "study lessons," which form the core of the Japanese practice of "Lesson Study." The authors describe the components of a research lesson and examine the impact that such lessons have in the improvement of Japanese education. They also consider what the key social, cultural, and political conditions are that support the lesson study system, in comparison with those of the United States.

Nelson, B. S. & Sassi, A. (2000) Shifting approaches to Supervision: The case of mathematics supervision, *Educational Administration Quarterly,* 36(3), 553–584.

These authors examine how supervisory practices have not taken account of subject-matter content but have focused primarily on pedagogical process. They address ways that administrators can better support standards-based instruction by shifting their approaches to supervision to attend to the intersection of process and content. The authors report on a study that looked at what administrators thought significant when viewing the same videotape of a fifth-grade mathematics lesson at the beginning and end of a professional development seminar on supervision. The authors conclude that administrators need to have adequate subject-matter knowledge for doing supervision and suggest several possible directions for achieving this shift.

Reitzug, Ulrich C. (1997) Images of Principal Instructional Leadership: From Super-Vision to Collaborative Inquiry. *Journal of Curriculum & Supervision.* vol. 12 no.4 (pp. 324–343.)

The author examines images of principals' instructional leadership, based on 10 supervision textbooks published between 1985 and 1995. He looks at how textbooks have portrayed principals as experts and superiors, teachers as deficient and voiceless, teaching as fixed technology, and supervision as a discrete intervention. Images of professional growth (that stress collegiality and continuous improvement) suggested by studies of successful schools differ significantly from these textbook images.

Schifter, D. (1999). Reasoning About Operations: Early algebraic thinking, grades K–6, in L. Stiff & F. Curio (Eds.) *Mathematical Reasoning, K–12: 1999 NCTM Yearbook* (pp. 62–81), Reston, VA: National Council of Teachers of Mathematics.

The author takes a close look at the development of students' operation sense in K–6 classrooms. She makes a case for the importance of organizing classrooms around students' mathematical ideas so that they can build understanding of the algebra they will encounter in later grades. The author describes a number of scenarios from elementary classrooms in which teachers are working to align their instructions with standards-based practice and which support students to articulate their mathematical ideas. She suggests that what can be learned about the development of students' operation sense from these scenarios can provide useful information about preparing young children for algebra.

Spillane, J., Halverson, R., & Diamond, J. (2001). Investigating school leadership practice: A distributed perspective. *Educational Researcher, 30*(3), 23–28.

These authors propose a framework for investigating how school leadership is "distributed" or stretched over the school's social and situational contexts. The authors ground their conceptualization of "distributed leadership" in activity theory and distributed cognition and lay out directions for a research program in which the nature of distributed leadership can be observed and analyzed.

Tracy, Saundra J. (1995). How historical concepts of supervision relate to supervisory practices today. *Clearing House. 68*(5), 320–325.

The author describes seven phases in the evolution of supervisory practice in the schools. She looks at each historical phase in relation to its purpose (assisting or assessing), focus or emphasis, the personnel typically involved, the skills needed to implement supervision, and the assumptions surrounding the process.

Books

Driscoll, Mark. (1999). *Fostering Algebraic thinking: A guide for teachers grades 6–10.* Portsmouth, NH: Heinemann.

This book is intended for teachers who want to reflect on their thinking about the teaching and learning of pre-algebra and early algebra. It builds on the move to base the teaching and learning of mathematics on explicit standards. The author presents a detailed description of his perspective on algebraic thinking and provides a framework for the use of classroom questions that foster the development of algebraic thinking in students from grades 6–12.

Elbow, Peter. (1986). *Embracing contraries: explorations in learning and teaching.* New York: Oxford University Press.

Mr. Elbow argues that what is actually most natural in teaching and learning is a rich messiness of paradox and contradiction. Consequently, he points out that we need to alter our view of how people learn and how teachers should teach and grade. In the book, he explores the learning process, the teaching process, the evaluation process, and the nature of inquiry.

Fosnot, C.T., and M. Dolk. (2001). *Young Mathematicians at Work: Constructing Number Sense, Addition, and Subtraction.* Portsmouth, NH: Heinemann

Fosnot, C.T., and M. Dolk. (2001). *Young Mathematicians at Work: Constructing Multiplication and Division.* Portsmouth, NH: Heinemann

Fosnot, C.T., and M. Dolk. (2001). *Young Mathematicians at Work: Constructing Fractions, Decimals, and Percents.* Portsmouth, NH: Heinemann

In these three books, the authors present students' approaches to solving mathematical problems, and they describe how teachers work to support and develop their students' mathematical thinking. Fosnot and Dolk lay out and explain the big ideas that comprise an understanding of the mathematical concepts addressed in each book.

Glickman, Carl D. (1992). *Supervision in Transition.* Washington, D.C.: Association for Supervision and Curriculum Development.

This collection of essays, written by leading scholars in the field of supervision, explore the connections between the current changes in school organization and governance structures and the supervisory skills and relationships that are called for by these changes. The collection provides an historical overview of the supervisory process, explorations of promising practices, and consideration of the preparation of teachers. The book acknowledges the need for supervision to entail professional inquiry and collegiality.

Glickman, C. D., Gordon, S. P., and Ross-Gordon, J. M. (1998) *Supervision of Instruction: A Developmental Approach.* Boston: Allyn and Bacon.

This textbook offers a "developmental approach" to the practice of supervision. It is built on the assumption that the aim of supervision is to help teachers develop as reflective, autonomous professionals and that supervision itself should ultimately be non-directive. The authors provide a variety of positions, ideas, and practices as well as theoretical grounding and case examples.

Sergiovanni, Thomas J. & Starratt, Robert J. (1993). *Supervision: a redefinition, fifth edition.* Boston, MA: McGraw-Hill.

The authors propose a reconceptualization of the supervisory role and its place within the school community. In this edition, they place professional and moral authority rather than bureaucratic authority as the central force behind what teachers should do and how supervision should be done. They replace the metaphor of organization with that of community to describe the nature of schools and schooling. While still emphasizing the importance of traditional supervisory skills and practices, the authors stress the moral relationship that needs to be established between supervisor and teachers for these skills to be exercised wisely.

Stigler, J. W., & Hiebert, J. (1999). *The teaching gap: Best ideas from the world's teachers for improving education in the classroom.* New York, NY: Summit Books.

Drawing on the conclusions of the Third International Mathematics and Science Study (TIMSS), Mr. Stigler and Hiebert offer an action plan for improving education in the United States. The TIMSS study used video to observe a large number of classroom lessons in the U.S., Japan, and Germany. In the book, the authors use data from the study, especially from eighth-grade mathematics classrooms, to examine why the quality of teaching in the U.S. lags behind that of our peers in other countries. Drawing particularly from the Japanese practices, including Japanese "lesson study," the authors argue that U.S. schools can be restructured as places where teachers can engage in career-long learning and classrooms can become laboratories for developing new teaching-centered ideas.

Curricula

Connected Mathematics Project (CMP) is a mathematics curriculum for grades 6–8. It was developed at Michigan State University with funding from the National Science Foundation, and it is published by Prentice Hall. The materials provide opportunities for students to investigate mathematics concepts that come from everyday situations.

Investigations in Number, Data, and Space is a K–5 mathematics curriculum developed at TERC in Cambridge MA with funding from the National Science Foundation. Investigations provides students with in-depth experiences in numbers, data, geometry, and the mathematics of change. It engages students in the exploration of major mathematical ideas and supports them to develop their own approaches to solving problems based on their knowledge and understanding of mathematical relationships. The curriculum also offers support for teachers in both mathematics content and pedagogy.

Videotapes

Mathematical Inquiry through Video: Tools for professional growth. A package of videos and teacher development materials developed by BBNT Solutions LLC.

This video package consists of ten video cases of middle school mathematics classrooms. They depict teachers who are working to change their teaching practice according to the NCTM standards and highlight the challenges, ideas, and issues these and other teachers face in this process. Each video is accompanied by a facilitator's guide which includes a description of the video; background information on the school, teacher, and classroom in the video; suggested workshop design features with video-related mathematical and pedagogical issues and activities; and a complete transcript. For more information on this series, you may contact BBN at 10 Moulton Street, Cambridge, MA. Or you may wish to speak with one of the authors of the materials, Fadia Harik, at f.harik@comcast.net or fadia.harik@umb.edu.

Mathematics: Assessing Understanding. A Series of Videotapes for Staff Development, featuring Marilyn Burns. While Plains, NY. Cuisenaire Company of America, Inc.

This series consists of three videotapes and accompanying teacher's discussion guide, showing a collection of individual assessments of mathematical understanding with students ages 7 through 12. All of the assessments address students' ability to estimate, reason numerically, and compute in problem-solving situations. The one-on-one interviews model for teachers the kinds of questions that are useful for gaining insights into how students are thinking and what they understand. Mathematical topics included are: number sense and the place value structure of our number system; estimation, numerical reasoning, and computation with whole numbers; and fractions.

Relearning to Teach Arithmetic. A series of videotapes for staff development, funded by the NSF and developed at TERC, Cambridge, MA. Published by Dale Seymour Publications, White Plains, NY: an imprint of Addison Wesley, Longman, Inc.

This video series provides teachers with a structured opportunity to explore how children develop facility with the four operations (addition, subtraction, multiplication, and division) and how teachers can foster the development of this facility. It consists of two packages, each containing 3 videos. The first package focuses on addition and subtraction and the second focuses on multiplication and division. Each package also includes a study guide that outlines six professional development sessions. During each session, teachers view and discuss segments of the tapes and work on related mathematics problems. Each package is adaptable to a variety of settings such as after-school sessions, release-day professional development sessions, or summer staff development experiences. The materials may also be integrated into a longer course or seminar on the teaching and learning of elementary mathematics.

Talking Mathematics. A professional development resource package funded by the NSF and developed at TERC, Cambridge, MA. Published by Heinemann, Portsmouth, NH.

This package includes a videotape program, a resource guide for staff developers and university instructors, and a book for teachers who are interested in supporting talk and mathematical inquiry in their classrooms. The goal of the package is to provide teachers and staff developers with resources that can help them cultivate good mathematical discourse. The video program consists of an introductory videotape, four twenty-minute videotapes on aspects of children's talk, six short classroom episodes, and a twenty-minute summary of a Talking Mathematics teacher seminar. The package can be adapted to a variety of professional development settings.

Teaching Math Video Libraries. A series of videotapes for staff development, funded by the Annenberg Foundation/Corporation for Public Broadcasting and produced by WGBH (1995). S. Burlington, VT: The Annenberg/CPB Math and Science Collection, (800) 965-7373.

This video series provides visual examples of standards-based teaching and learning. Four Teaching Math "libraries" are available: K–4, 5–8, 9–12, and a K–12 assessment library. The grade level libraries each consist of a set of content standard videos, a set of process standard videos, and guidebooks. The assessment library provides case study videos that examine assessment issues in two different classes and a sequence of vignettes that show a variety of assessment techniques from several classes. This collection of video libraries provides grounded images of what classrooms may look like when teachers are developing their teaching in accordance with NCTM standards.

NOTES

Notes

Notes

NOTES

Notes

Notes

Notes